Sound of 2 Black Holes Colliding

By

Dr. Jon Schiller, PhD

Sound of 2 Black Holes Colliding

All rights reserved. No part of this book may be reproduced in any form without written permission from the author, except by a reviewer who may quote brief passages in a review to be printed in a newspaper, a magazine or a website.

WRITTEN BY JON SCHILLER, PhD
Printed in the United States of America
2016
First Printing

ISBN-13: 978-1534701328
JON SCHILLER SOFTWARE
jonsch1@verizon.net
http://www.jonschilleroptions.com/
© 2016 by Emilie M. Smyth

Sound of 2 Black Holes Colliding

By

Dr. Jon Schiller, PhD

2016

Sound of 2 Black Holes Colliding

Dr. Jon Schiller's books

Options Trading Books
Insider's Automatic Options Strategy
Self-Adaptive Options & Currency Trading
The 100% Return Options Trading Strategy
Avoid Financial Fraud by Using Weekly Options
Weekly Options Trading to Maximize Your Capital
Options Profits Using Decision Charts Using Strategies Developed over 2 Decades of Options Trading
Weekly OEX Options to Grow Trading Capital Rapidly
OEX Weekly Trading Tips & Newsletter Compilations
Double Your Money with Weekly Options Condors
Grow Your Wealth Using Weekly Options Trading
9 Weeks of Trading Weekly Options
Options Strategy to Profit during Extreme Volatility
Options Profits using Sustainable Energy Companies
Weekly Index Options Trading Tips to Increase Profits
Weekly Options Profits Using Tablet Computer
Trade Weekly Options Using Android Mobile Devices
Profit when Algorithmic Trading Systems Cause Flashcrashes
Mobile Devices Revolution
Weekly Index Options Impacted by Washington Programs
Weekly Options Trading Algorithm using 2sig & WWI
Win with Weekly Options
Weekly Options in 2015
Weekly Options for Monthly Income

Jon Schiller's Documentary & Technical Books
21st Century Cosmology
Quantum Computers,
Life Style to Extend Life Span
Visual Basic Express & Java
Internet View of the Arabic World
Bullet Trains Go Over 365 mph

Sound of 2 Black Holes Colliding

Global Change & Energy Policy
Human Evolution: Neanderthals & Homosapiens
Nano Technology Developments
Big Bang and Black Holes
Avoid Terrorist Attacks
Cyber Attacks & Protection
US Government Debt Story
Education in the 21st Century
Prostate Cancer
Chemical Bonding
Genome Mapping
Bin Laden Demise as Seen on Internet in 2011
Making Movies using PowerPoint, CamStudio,
    WinMovieMaker & WinDVDMaker
Will Barak Obama Win 2012 Presidential Elections?
Wikipedia & Free Speech
History of Civilization
Walking to Improve Health
Great Technology Transitions of Civilization
Higgs Boson Particle and Impact on Cosmology
Mars Mission and Shuttle History & Replacement
Political Finger Pointing Impact on Wall Street
MidEast Revolutions, Iraq, Afghanistan, Egypt, Libya,
    Syria, Tunisia, Yemen, Israel vs Enemies.
Nanorobots & Microrobots   Exciting Tools of Future
GPS Technology for Walking, Driving, Boating, Flying
DNA Chemistry: DNA Damage & Repair; Aid to Human
    Health
Ruby on Rails for Creating Interactive Websites
Dreamliner
Dark Matter Evidence found at CERN
Optogenetics Research
Washington Political Battles
Global Education for Universities and Colleges
Spy Whistleblower
Intelligent Drones Future for Military Aircraft

Sound of 2 Black Holes Colliding

Fly by Wire Aircraft, Fighters, Drones, and Airliners
How to Grow Old Gracefully
Wealth Accumulating to the Few
Revising US Constitution to make it Suitable for Now
Open Heart Bypass Surgery to Fix Defects
F-35 Lightning II
Russian Cultural Tour-Saint Petersburg and Moscow
Anti-Aging Research   Live Much Longer
Aerospace Wonder   Stealth Aircraft
Brain Study 2015
Private Flying in Small Aircraft
How to Eliminate ISIS
Around the World Trip in a LearJet
String Theory
Avoid US Downfall

Fiction Novels & True Stories
IBEX
Lost in Space
Masada Never Again
Multihulls
Ultra Taiwan Fighter
Irrational Indictment & Imprisonment
Family History of a Successful Aerospace Executive
Six Gay Love Tales
Six Gay Love Tales, Vol.2
Five Gay Love Conversions
Gay Love Techniques

Sound of 2 Black Holes Colliding

Table of Contents

Introduction		ix
Chapter 1	Gravitational Waves from Black Holes Colliding	1
Chapter 2	Artists' Conceptions of Black Holes Colliding	33
Chapter 3	Black Hole Description	37
Chapter 4	Sound of Two Black Holes Colliding	47
Chapter 5	What Happens When 2 Black Holes Collide	61
Chapter 6	The Particle That Broke a Cosmic Speed Limit	83
Chapter 7	From Einstein's Theory to Gravity's Chirp	83
Chapter 8	Listen to the Collision of Two Black Holes Einstein Was Right	97
Chapter 9	Why Black Holes Have a Hard Time Getting Together	107
Appendix A	Glossary	113

Sound of 2 Black Holes Colliding

Introduction

I was made aware of the subject of this book while attending the Caltech Alumni Seminar on 21 May 2016 on campus. I missed the Seminar lecture but the subject sounded so interesting I decided to write a book about it in order to learn more about it.

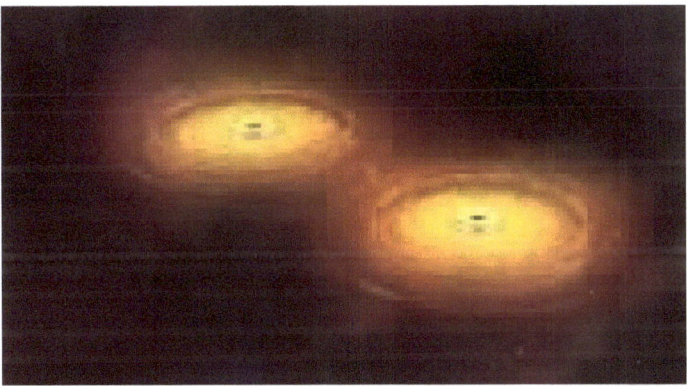

(Above) Image of 2 Back Holes close to colliding taken from the Internet. The image above was made using the Hubble Space Telescope.

On 14 September 2015 there was a collision of black holes.

It has been speculated for some time that two black holes can collide. Once they came so close that they could not escape each other's gravity, they could merge to become one bigger black hole. Even when simulating this event on powerful computers,

Sound of 2 Black Holes Colliding

we could not fully understand it. However, we did know that a black hole merger would produce tremendous energy and send massive ripples through the space time fabric of the Universe. These ripples are called gravitational waves.

We have known of galaxies in which two super-massive black holes have moved dangerously close to each other. Theoretical models predicted that these black holes would spiral toward each other until they would eventually collide.

Gravitational waves have been a fundamental prediction of Einstein's theory of general relativity. Detecting them provides an important test of our understanding of gravity. It also provides important new insights into the physics of black holes. Large instruments capable of detecting gravitational waves from outer space have been built in recent years. Even more powerful instruments are under construction.

Chapter 1. Gravitational Waves from Black Holes Colliding

LIGO Founders Receive Prestigious Kavli Prize in Astrophysics

The 2016 Kavli Prize in Astrophysics has been awarded to the three founders of the Laser Interferometer Gravitationla Wave Observatory (LIGO): Caltech's Ronald W.P.Drever, professor of physics, emeritus, and Kip S. Thorne (BS '62), the Richard P. Feynman Professor od Theoretical Physics, Emeritus; and MIT's Rainer Weiss, professor of physics, emeritus.

The $1 million prize, presented once every two years, honors the three for teir instrumental role in establishing LIGO, an effort that led to the direct detection of gravitational waves (which are ripples in the fabric of space and time predicted a century earlier by Albert Einstein's General Theory of Relativity). On February

Sound of 2 Black Holes Colliding

11, 2016, the international LIGO team announced the first observation of gravitational waves arriving at Earth.

The **Laser Interferometer Gravitational Wave Observatory (LIGO)** is a large-scale physics experiment and observatory to detect gravitational waves. LIGO is a joint project among scientists from several colleges and universities. Scientists involved in the project and the analysis of the data for gravitational wave astronomy are organized by the LIGO Scientific Collaboration which includes more than 900 scientists worldwide, as well as 44,000 active Einstein@Home users. LIGO is funded by the National Science Foundation (NSF), with important contributions from the UK Science and Technology Facilities Council, the Max Planck Society of Germany, and the Australian Research Council. LIGO is the largest and most ambitious project ever funded by the NSF.

History

The LIGO concept is built upon early work by many scientists to test a component of Albert Einstein's theory of relativity, the existence of gravitational waves. Starting in the 1960s, American scientists, including Joseph Weber, as well as Soviet scientists Mikhail Gertsenshtein and Vladislav Pustovoit, conceived of basic ideas and prototypes of laser interferometry, and in 1967 Rainer Weiss of MIT published an analysis of interferometer use and initiated the construction of a prototype with military funding, but it was terminated before it could become operational. Starting in 1968, Kip Thorne initiated theoretical efforts on gravitational waves and their sources at Caltech, and

Sound of 2 Black Holes Colliding

was convinced that gravitational wave detection would eventually succeed.

Prototype interferometric gravitational wave detectors (interferometers) were built in the late 1960s by Robert L. Forward and colleagues at Hughes Research Laboratories (with mirrors mounted on a vibration isolated plate rather than free swinging), and in the 1970s (with free swinging mirrors between which light bounced many times) by Weiss at MIT, and then by Heinz Billing and colleagues in Garching, Germany, and then by Ronald Drever, James Hough and colleagues in Glasgow, Scotland.

In 1980, the NSF funded the study of a large interferometer led by MIT (Paul Linsay, Peter Saulson, Rainer Weiss), and the following year, Caltech constructed a 40-meter prototype (Ronald Drever and Stan Whitcomb). The MIT study established the feasibility of interferometers at a 1-kilometer scale with adequate sensitivity.

Under pressure from the NSF, MIT and Caltech were asked to join forces to lead a LIGO project based on the MIT Study and on experimental work at Caltech, MIT, Glasgow, and Garching. Drever, Thorne, and Weiss formed a LIGO steering committee, though they were turned down for funding in 1984 and 1985. By 1986, they were asked to disband the steering committee and a single director, Rochus E. Vogt, was appointed. In 1988, a research and development proposal achieved funding.

From 1989 through 1994, LIGO failed to progress technically and organizationally. Only political efforts continued to acquire

Sound of 2 Black Holes Colliding

funding. Ongoing funding was routinely rejected until 1991, when the US Congress agreed to fund LIGO for the first year for $23 million. However, requirements for receiving the funding were not met or approved, and the NSF questioned the technological and organizational basis of the project. By 1992, LIGO was restructured with Drever no longer a direct participant. Ongoing project management issues and technical concerns were revealed in NSF reviews of the project, resulting in the withholding of funds until they formally froze spending in 1993.

In 1994, after consultation between relevant NSF personnel, LIGO's scientific leaders, and the presidents of MIT and Caltech, Vogt stepped down and Barry Barish (Caltech) was appointed laboratory director, and the NSF made clear that LIGO had one last chance for support. Barish's team created a new study, budget, and project plan with a budget exceeding the previous proposals by 40%. Barish proposed to the NSF and National Science Board to build LIGO as an evolutionary detector, where detection of gravitational waves with the Initial LIGO would be possible, and with advanced LIGO would be probable. This new proposal received NSF funding, Barish was appointed Principal Investigator, and the increase was approved. In 1994, with a budget of USD 395 million, LIGO stood as the largest overall funded NSF project in history. The project broke ground in Hanford, Washington, in late 1994 and in Livingston, Louisiana, in 1995. As construction neared completion in 1997, under Barish's leadership two organizational institutions were formed: LIGO Laboratory and LIGO Scientific Collaboration (LSC). The LIGO laboratory consisted of the facilities supported by the NSF under LIGO Operation and Advanced R&D; this included

Sound of 2 Black Holes Colliding

administration of the LIGO detector and test facilities. The LIGO Scientific Collaboration is a forum for organizing technical and scientific research in LIGO. It is a separate organization from LIGO Laboratory with its own oversight. Barish appointed Weiss as the first spokesperson for this scientific collaboration.

The Initial LIGO operations between 2002 and 2010 did not detect any gravitational waves. In 2004, under Barish, the funding and groundwork were laid for the next phase of LIGO development (called "Enhanced LIGO"). This was followed by a multi-year shut-down while the detectors were replaced by much improved "Advanced LIGO" versions. Much of the research and development work for the advanced LIGO machines was based on pioneering work for the GEO600 detector at Hannover, Germany. By February 2015, the detectors were brought into engineering mode in both locations.

By mid-September 2015 "the world's largest gravitational-wave facility" completed a 5-year US$200-million overhaul at a total cost of $620 million. On September 18, 2015, Advanced LIGO began its first formal scientific observations at about four times the sensitivity of the Initial LIGO interferometers. Its sensitivity will be further enhanced until it reaches the design sensitivity specified in the enhanced concept around 2021.

On February 11, 2016, the LIGO Scientific Collaboration and Virgo Collaboration published a paper about the detection of gravitational waves, from a signal detected at 09.51 UTC (Universal Time Coordinated) on 14 September 2015 of two ~30

Sound of 2 Black Holes Colliding

solar mass black holes merging about 1.3 billion light years from Earth.

Current executive director David Reitze (California Institute of Technology and University of Florida) announced the findings at a media event in Washington D.C. while executive director emeritus Barry Barish (Caltech) presented the first scientific paper of the findings at CERN (in Switzerland) `to the physics community.

On May 2, 2016, members of the LIGO Scientific Collaboration and other contributors were awarded a Special Breakthrough Prize in Fundamental Physics for contributing to the direct detection of gravitational waves.

LIGO's mission is to directly observe gravitational waves of cosmic origin. These waves were first predicted by Einstein's General Theory of Relativity in 1916, when the technology necessary for their detection did not yet exist. Their existence was indirectly confirmed when observations of the binary pulsar PSR 1913+16 in 1974 showed an orbital decay which matched Einstein's predictions of energy loss by gravitational radiation. The Nobel Prize in Physics in 1993 was awarded jointly to Russell A. Hulse and Joseph H. Taylor Jr. "for the discovery of a new type of pulsar, a discovery that has opened up new possibilities for the study of gravitation, including gravitational waves, called GW."

Direct detection of gravitational waves has long been sought. Their discovery would launch a new branch of astronomy to complement electromagnetic telescopes and neutrino

Sound of 2 Black Holes Colliding

observatories. Joseph Weber pioneered the effort to detect gravitational waves in the 1960s through his work on mass resonant bar detectors, also known as the *Weber bar*. Bar detectors continue to be used at six sites worldwide. By the 1970s, scientists including Rainer Weiss realized the applicability of laser interferometry to gravitational wave measurements. Robert Forward operated an interferometric detector at Hughes Aircraft Company in the early 1970s.

As early as the 1960s, and perhaps before that, there were papers published on wave resonance of light and gravitational waves. Works were published in 1971 on methods to exploit this resonance for the detection of high-frequency gravitational waves. In 1962, M. E. Gertsenshtein and V. I. Pustovoit published the very first paper describing the principles for using interferometers for the detection of very long wavelength gravitational waves. The authors argued that by using interferometers the sensitivity can be 10^7 to 10^{10} times better than by using electromechanical experiments. Later, in 1965, Braginsky extensively discussed gravitational-wave sources and their possible detection. He pointed out the 1962 paper and mentioned the possibility of detecting gravitational waves if the interferometric technology and measuring techniques improved.

In August 2002, LIGO began its search for cosmic gravitational waves. Measurable emissions of gravitational waves were expected from binary systems (collisions and coalescences of neutron stars or black holes), supernova explosions of massive stars (which form neutron stars and black holes), accreting neutron stars, rotations of neutron stars with deformed crusts, and

Sound of 2 Black Holes Colliding

the remnants of gravitational radiation created by the birth of the universe. In theory, a person can observe more exotic hypothetical phenomena, such as gravitational waves caused by oscillating cosmic strings (hypothetical one dimensional topological defects which may have formed during a symmetry breaking phase transition in the early universe) or by colliding domain walls.

Since the early 1990s, physicists have thought that technology hasd evolved to the point where detection of gravitational waves of significant astrophysical interest was possible.

Observatories

LIGO operates two gravitational wave observatories in unison: the first is the LIGO Livingston Observatory (30°33'46.42"N 90°46'27.27"W) in Livingston, Louisiana, and the second is the LIGO Hanford Observatory, on the DOE (US Department of Energy) Hanford Site (46°27'18.52"N 119°24'27.56"W), located near Richland, Washington. These sites are separated by 3,002 kilometers (1,865 miles). See small US map in the image on the next page.

Since gravitational waves are expected to travel at the speed of light, this distance corresponds to a difference in gravitational wave arrival times of up to ten milliseconds. Through the use of trilateration, the difference in arrival times helps to determine the source of the wave.

Sound of 2 Black Holes Colliding

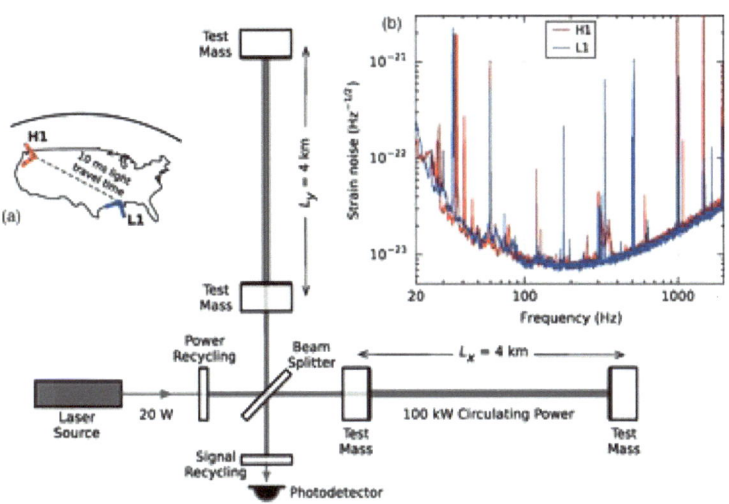

(above) Diagram of Interferometer set-up at each LIGO site.

Each observatory supports an L-shaped ultra high vacuum system, measuring 4 kilometers (2.5 miles) on each side. Up to five interferometers can be set up in each vacuum system.

The LIGO Livingston Observatory houses one laser interferometer in the primary configuration. This interferometer was successfully upgraded in 2004 with an active vibration isolation system based on hydraulic actuators providing a factor of 10 isolation in the 0.1–5 Hz band. Seismic vibration in this band is chiefly due to micro-seismic waves and anthropogenic sources (traffic, logging, etc.).

The LIGO Hanford Observatory houses one interferometer, almost identical to the one at the Livingston Observatory. During the Initial and Enhanced LIGO phases, a half-length interferometer operated in parallel with the main interferometer. For this 2 km interferometer, the Fabry-Pérot arm cavities had the same optical finesse, and thus half the storage time, as the 4 km interferometers. With half the storage time, the theoretical strain sensitivity was as good as the full length interferometers above 200 Hz but only half as good at low frequencies. During the same era, Hanford retained its original passive seismic isolation system due to limited geologic activity in Southeastern Washington.

Operation

The parameters in this section refer to the Advanced LIGO experiment. The primary interferometer consists of two beam lines 4 km long which form a power-recycled Michelson interferometer with Gires-Tournois etalon arms. A pre-stabilized laser emits a beam that passes through a power recycling mirror. The mirror fully transmits light incident from the laser and reflects light from the other side increasing the power of the light field between the mirror and the subsequent beam splitter. From the beam splitter the light travels along two orthogonal arms. By the use of partially reflecting mirrors, Fabry-Pérot cavities are created in both arms that increase the effective path length of laser light in the arm. The power of the light field in the cavity is 100 kilowatts.

When a gravitational wave passes through the interferometer, the space-time in the local area is altered. Depending on the source of the wave and its polarization, this results in an effective change in length of one or both of the cavities. The effective

Sound of 2 Black Holes Colliding

length change between the beams will cause the light currently in the cavity to become very slightly out of phases (antiphase) with the incoming light. The cavity will therefore periodically get very slightly out of coherence and the beams, which are tuned to destructively interfere at the detector, will have a very slight periodically varying detuning. This results in a measurable signal.

After an equivalent of approximately 280 trips down the 4 km length to the far mirrors and back again, the two separate beams leave the arms and recombine at the beam splitter. The beams returning from two arms are kept out of phase so that when the arms are both in coherence and interference (as when there is no gravitational wave passing through), their light waves subtract, and no light should arrive at the photodiode. When a gravitational wave passes through the interferometer, the distances along the arms of the interferometer are shortened and lengthened, causing the beams to become slightly less out of antiphase. This results in the beams coming in phase, creating a resonance, hence, some light arrives at the photodiode, indicating a signal. Light that does not contain a signal is returned to the interferometer using a power recycling mirror, thus increasing the power of the light in the arms. In actual operation, noise sources can cause movement in the optics which produces similar effects to real gravitational wave signals; a great deal of the art and complexity in the instrument is in finding ways to reduce these spurious motions of the mirrors. Observers compare signals from both sites to reduce the effects of noise.

Sound of 2 Black Holes Colliding

An interesting fact is that the mechanism for the operation of lasers – stimulated emission - was also predicted by Einstein in 1917 and then realized by experiments after more than four decades.

Observations

Based on current models of astronomical events, and the predictions of the General Theory of Relativity, gravitational waves that originate tens of millions of light years from Earth are expected to distort the 4 kilometer mirror spacing by about 10^{-18} m, less than one-thousandth the charge diameter of a proton. Equivalently, this is a relative change in distance of approximately one part in 10^{21}. A typical event which might cause a detection event would be the late stage in-spiral and merger of two 10 solar mass black holes, not necessarily located in the Milky Way galaxy, which is expected to result in a very specific sequence of signals often summarized by the slogan "chirp, burst, quasi-normal mode ringing, exponential decay."

In their fourth Science Run at the end of 2004, the LIGO detectors demonstrated sensitivities in measuring these displacements to within a factor of 2 of their design.

During LIGO's fifth Science Run in November 2005, sensitivity reached the primary design specification of a detectable strain of one part in 10^{21} over a 100 Hertz bandwidth. The baseline in-spiral of two roughly solar-mass neutron stars is typically expected to be observable if it occurs within about 8 million parsecs (26×10^6 light year), or the vicinity of the Local Group,

Sound of 2 Black Holes Colliding

averaged over all directions and polarizations. Also at this time, LIGO and GEO600 (the German-UK interferometric detector) began a joint science run, during which they collected data for several months. Virgo (the French-Italian interferometric detector) joined in May 2007. The fifth science run ended in 2007, after extensive analysis of data from this run did not uncover any unambiguous detection events.

In February 2007, GRB 070201, a short gamma ray arrived at Earth from the direction of the Andromeda Galaxy. The prevailing explanation of most short gamma ray bursts is the merger of a neutron star with either a neutron star or a black hole. LIGO reported a non-detection for GRB 070201, ruling out a merger at the distance of Andromeda with high confidence. Such a constraint was predicated on LIGO eventually demonstrating a direct detection of gravitational waves.

On 11 February 2016, the LIGO and Virgo collaborations announced the first observation of a gravitational wave. The signal was named GW150914. The waveform showed up on 14 September 2015, two days from when the Advanced LIGO detectors started collecting data after their upgrade. It matched the predictions of general relativity for the inward spiral and merger of a pair of black holes and the subsequent 'ringdown' of the resulting single black hole. The observations demonstrated the existence of binary stellar-mass black hole systems and the first observation of a binary black hole merger.

Sound of 2 Black Holes Colliding

Western leg of LIGO interferometer on Hanford Reservation (image above).

Northern leg (x-arm) of LIGO interferometer on Hanford Reservation (image above).

After the completion of Science Run 5, the Initial LIGO was upgraded with certain technologies that resulted in an improved-performance configuration dubbed Enhanced LIGO. Some of the improvements in Enhanced LIGO included:

- Increased laser power
- Homodyne detection
- Output mode cleaner
- In-vacuum readout hardware

Sound of 2 Black Holes Colliding

Science Run 6 (S6) began in July 2009 with the enhanced configurations on the 4 km detectors. It concluded in October 2010, and the disassembling of the original detectors began. By mid-September 2015, LIGO Scientific Collaboration included more than 900 scientists worldwide.

Advanced LIGO

The LIGO Laboratory, funded by the National Science Foundation with contributions from the GEO600 Collaboration and ANU (Australian National University) and Adelaide Universities in Australia, and with participation by the LIGO Scientific Collaboration, has installed the new Advanced LIGO detectors in the LIGO Observatory infrastructures. This new detector is designed to improve the sensitivity of the Initial LIGO by more than a factor of 10 once fully commissioned.

The LIGO Laboratory started the first observing run 'O1' with the Advanced LIGO detectors in September 2015 at a sensitivity roughly 4 times greater than the Initial LIGO for some classes of sources (e.g., neutron-star binaries), and a much greater sensitivity for larger systems with their peak radiation at lower audio frequencies. Further observing runs will be interleaved with commissioning efforts to further improve the sensitivity. It is aimed to achieve design sensitivity in 2021.

Future

LIGO-India, or Indigo, is a planned collaborative project between the LIGO Laboratory and the Indian Initiative in Gravitational-wave Observations (IndIGO) to create a world-class gravitational

Sound of 2 Black Holes Colliding

wave detector in India. The LIGO Laboratory, in collaboration with the US National Science Foundation and Advanced LIGO partners from the U.K., Germany and Australia, has offered to provide all of the designs and hardware for one of the three planned Advanced LIGO detectors to be installed, commissioned, and operated by an Indian team of scientists in a facility to be built in India.

The expansion of worldwide activities in gravitational-wave detection to produce an effective global network has been a goal of LIGO for many years. In 2010, a developmental roadmap issued by the Gravitational Wave International Committee (GWIC) recommended that an expansion of the global array of interferometric detectors be pursued as a highest priority. Such a network would afford astrophysicists with more robust search capabilities and higher scientific yields. The current agreement between the LIGO Scientific Collaboration and the Virgo collaboration links three detectors of comparable sensitivity and forms the core of this international network. Studies indicate that the localization of sources by a network that includes a detector in India would provide significant improvements. Improvements in localization averages are predicted to be approximately an order of magnitude, with substantially larger improvements in certain regions of the sky.

The NSF was willing to permit this relocation, and its consequent schedule delays, as long as it did not increase the LIGO budget. Thus, all costs required to build a laboratory equivalent to the LIGO sites to house the detector would have to be borne by the host country. The first potential distant location was at AIGO in

Sound of 2 Black Holes Colliding

Western Australia. However, the Australian government was unwilling to commit funding by the 1 October 2011 deadline.

A location in India was discussed at a Joint Commission meeting between India and the US in June 2012. In parallel, the proposal was evaluated by LIGO's funding agency, the NSF. As the basis of the LIGO-India project entails the transfer of one of LIGO's detectors to India, the plan would affect work and scheduling on the Advanced LIGO upgrades already underway. In August 2012, the U.S. National Science Board approved the LIGO Laboratory's request to modify the scope of Advanced LIGO by not installing the Hanford "H2" interferometer, and to prepare it instead for storage in anticipation of sending it to LIGO-India. In India, the project was presented to the Department of Atomic Energy and the Department of Science and Technology for approval and funding. On 17 February 2016, less than a week after LIGO's landmark announcement about the detection of gravitational waves, Indian Prime Minister Narendra Modi announced that the Cabinet has granted 'in-principle' approval to the LIGO-India mega science proposal.

Einstein Telescope

The Einstein Telescope (ET) or Einstein Observatory, is a proposed third-generation ground-based gravitational wave detector, currently under study by some institutions in the European Union. It will be able to test Einstein's Theory of General Relativity in strong field conditions and realize precision gravitational wave astronomy.

Sound of 2 Black Holes Colliding

The ET is a design study project supported by the European Commission under the Framework Programme 7 (FP7) (using British spelling). It concerns the study and the conceptual design for a new research infrastructure in the emergent field of gravitational wave astronomy. See image below.

Motivation

The evolution of the current gravitational wave detectors Virgo and LIGO, as *first generation* detectors, is well defined. After the current upgrade to their so-called enhanced level, the detectors are evolving toward their second generation: the advanced Virgo and LIGO detectors. LIGO already detected gravitational waves in 2015 and further sensitivity upgrades promise many more detections to follow. But the sensitivity needed to test Einstein's theory of gravity in strong field conditions or to realize a precision gravitational wave astronomy, mainly of massive stellar bodies or of highly asymmetric (in mass) binary stellar systems goes beyond the expected performances of the advanced detectors and of their subsequent upgrades. For example, the fundamental limitations at low frequency of the sensitivity of the second generation detectors are given by the seismic noise, the related gravitational gradient noise (so-called Newtonian noise) and the thermal noise of the suspension last stage and of the test masses.

To circumvent these limitations new infrastructures are necessary: an underground site for the detector, to limit the effect of the seismic noise, and cryogenic facilities to cool down the mirrors to directly reduce the thermal vibration of the test masses.

Technical Groups

The ET-FP7 project, through its four technical working groups is addressing the basic questions in the realization of this proposed observatory: site location and characteristics (WP1), suspension design and technologies (WP2), detector topology and geometry (WP3), detection capabilities requirements and astrophysics potentialities (WP4).

Participants

The Einstein Telescope has been proposed by eight European research institutes (in local language spelling):

- European Gravitational Obsevatory
- Istituto Nazionale di Fisica Nucleare
- Max Planck Society
- Centre National de la Recherche Scientifique
- University of Birmingham
- University of Glascow
- NIKHEF
- Cardiff University

Sound of 2 Black Holes Colliding

KAGRA

The Kamioka Gravitational Wave Detector (KAGRA), formerly the Large Scale Cryogenic Gravitational Wave Telescope (LCGT), is a project of the gravitational wave studies group at the Institute for Cosmic Ray Research (ICRR) of the University of Tokyo. The ICRR was established in 1976 for cosmic ray studies, and is currently working on TAMA 300. The LCGT project was approved on 22 June 2010. In January 2012, it was given its new name, KAGRA, deriving the "KA" from its location at the Kamioka mine and "GRA" from gravity and gravitational radiation.

KAGRA has two sets of 3 km (1.9 mi) arm length laser interferometric gravitational wave detectors which were being built in the tunnels of Kamioka mine in Japan, sharing the mountain's interior with its neighbor, Kamioka Observatory. The excavation phase of tunnels was completed on 31 March 2014. KAGRA will detect chirp waves from binary neutron star coalescence at 240 Mpc (megaparse) away with a S/N (Signal-to-noise ratio) of 10. The expected number of detectable events in a year is two or three. To achieve the required sensitivity, several advanced techniques will be employed such as a low-frequency vibration-isolation system, a suspension point interferometer, cryogenic mirrors, a resonant side band extraction method, a high-power laser system and so on. KAGRA was initially hoped to begin operations in 2009 but is now likely to enter operation in 2018.

Sound of 2 Black Holes Colliding

Current Design of the Einstein Telescope

Although still in the early design study phase, the basic parameters are established.

Like KAGRA, it will be located underground to reduce seismic noise and "gravity gradient noise" caused by nearby moving objects.

The arms will be 10 km long (compared to 4 km for LIGO, and 3 km for Virgo and KAGRA), and like LISA, there will be three arms in an equilateral triangle, with two detectors in each corner.

In order to measure the polarization of incoming gravitational waves and avoid having an orientation to which the detector is insensitive, a minimum of two detectors are required. While this could be done with two 90° interferometers at 45° to each other, the triangular form allows the arms to be shared. The 60° arm angle reduces the interferometer's sensitivity, but that is made up for by the third detector, and the additional redundancy provides a useful cross-check.

Each of the three detectors would be composed of two interferometers, one optimized for operation below 30 Hz and one optimized for operation at higher frequencies.

The low-frequency interferometers (1 to 250 Hz) will use optics cooled to 10 K (−441.7 °F; −263.1 °C), with a beam power of about 18 kW in each arm cavity. The high-frequency ones (10 Hz to 10 kHz) will use room-temperature optics and a much higher recirculating beam power of 3 MW.

Tests of General Relativity

At its introduction in 1915, the General Theory of Relativity did not have a solid empirical foundation. It was known that it correctly accounted for the "anomalous" precession of the perihelion of Mercury and on philosophical grounds it was considered satisfying that it was able to unify Newton's Law with special relativity. That light appeared to bend in gravitational fields in line with the predictions of general relativity was found in 1919 but it was not until a program of precision tests was started in 1959 that the various predictions of general relativity were tested to any further degree of accuracy in the weak gravitational field limit, severely limiting possible deviations from the theory. Beginning in 1974 Hulse, Taylor and others have studied the behavior of binary pulsars experiencing much stronger gravitational fields than those found in the Solar System. Both in the weak field limit (as in the Solar System) and with the stronger fields present in systems of binary pulsars the predictions of general relativity have been extremely well tested.

The very strong gravitational fields that are present close to black holes, especially those supermassive black holes which are thought to power active galactic nuclei and the more active quasars, belong to a field of intense active research. Observations of these quasars and active galactic nuclei are difficult, and interpretation of the observations is heavily dependent upon astrophysical models other than general relativity or competing fundamental theories of gravitation, but they are qualitatively consistent with the black hole concept as modelled in general relativity. As a consequence of the equivalence principle, Lorentz Invariance holds locally in non-rotating, freely falling reference frames. Experiments related to Lorentz Invariance and thus to special relativity (that is, when gravitational effects can be neglected) are described in Tests of special relativity. In February 2016, the Advanced LIGO team announced that they had directly detected gravitational waves from a black hole merger.

Classical Tests of Special Relativity

Albert Einstein proposed three tests of general relativity, subsequently called the classical tests of general relativity, in 1916:

1. the perihelion precession of Mercury's orbit
2. the deflection of light by the Sun
3. the gravitational redshift of light

In the letter to the London *Times* on November 28, 1919, he described the theory of relativity and thanked his English colleagues for their understanding and testing of his work. He also mentioned three classical tests with comments: "The chief attraction of the theory lies in its logical completeness. If a single one of the conclusions drawn from it proves wrong, it must be given up; to modify it without destroying the whole structure seems to be impossible."

Perihelion Precession of Mercury

Under Newtonian physics, a two-body system consisting of a lone object orbiting a spherical mass would trace out an ellipse

Sound of 2 Black Holes Colliding

with the spherical mass at a focus. The point of closest approach, called the periapsis (or, because the central body in the Solar System is the Sun, perihelion), is fixed. A number of effects in the Solar System cause the perihelia of planets to precess (rotate) around the Sun. The principal cause is the presence of other planets which perturb one another's orbit. Another (much less significant) effect is solar oblateness.

Mercury deviates from the precession predicted from these Newtonian effects. This anomalous rate of precession of the perihelion of Mercury's orbit was first recognized in 1859 as a problem in celestial mechanics, by Urbain Le Verrier. His reanalysis of available timed observations of transits of Mercury over the Sun's disk from 1697 to 1848 showed that the actual rate of the precession disagreed from that predicted from Newton's theory by 38" (arc seconds) per tropical century (later re-estimated at 43"). A number of *ad hoc* and ultimately unsuccessful solutions were proposed, but they tended to introduce more problems. In general relativity, this remaining precession, or change of orientation of the orbital ellipse within its orbital plane, is explained by gravitation being mediated by the curvature of spacetime. Einstein showed that general relativity agrees closely with the observed amount of perihelion shift. This was a powerful factor motivating the adoption of general relativity.

Although earlier measurements of planetary orbits were made using conventional telescopes, more accurate measurements are now made with radar. The total observed precession of Mercury is 574.10±0.65 arc-seconds per century relative to the inertial ICFR. This precession can be attributed to the following causes:

Sources of the precession of perihelion for Mercury

Amount (arcsec/Julian century)	Cause
531.63 ±0.69[6]	Gravitational tugs of the other planets
0.0254	Oblateness of the Sun (quadrupole moment)
42.98 ±0.04[7]	General relativity
574.64±0.69	Total
574.10±0.65[6]	Observed

The correction by 42.98" is 3/2 multiple of classical prediction with PPN parameters $\gamma = \beta = 0$.

Thus the effect can be fully explained by general relativity. Recent calculations based on more precise measurements have not materially changed the situation.

The other planets experience perihelion shifts as well, but, since they are farther from the Sun and have longer periods, their shifts are lower, and could not be observed accurately until long after Mercury's. For example, the perihelion shift of Earth's orbit due to general relativity is 3.84 seconds of arc per century, and Venus's is 8.62". Both values are in good agreement with observation. The periapsis shift of binary pulsar systems have been measured, with PSR 1913+16 amounting to 4.2° per year. These observations are consistent with general relativity. It is also possible to measure periapsis shift in binary star systems which do not contain ultra-dense stars, but it is more difficult to model the classical effects precisely - for example, the alignment of the stars' spin to their orbital plane needs to be known and is hard to measure directly - so a few systems such as DI Herculis have been considered as problematic cases for general relativity.

In general relativity the perihelion shift σ, expressed in radians per revolution, is approximately given by:

$$\sigma = \frac{24\pi^3 L^2}{T^2 c^2 (1-e^2)},$$

Where L is the semi-major axis, T is the orbital period, c is the speed of light, and e is the orbital eccentricity.

Deflection of Light by the Sun

Henry Cavendish in 1784 (in an unpublished manuscript) and Johann Georg von Soldner in 1801 (published in 1804) had pointed out that Newtonian gravity predicts that starlight will bend around a massive object. The same value as Soldner's was calculated by Einstein in 1911 based on the equivalence principle alone. However, Einstein noted in 1915 in the process of completing general relativity, that his (and thus Soldner's) 1911 result is only half of the correct value. Einstein became the first to calculate the correct value for light bending.

Sound of 2 Black Holes Colliding

The first observation of light deflection was performed by noting the change in position of stars as they passed near the Sun on the celestial sphere. The observations were performed in May 1919 by Arthur Eddington and his collaborators during a total solar eclipse, so that the stars near the Sun (at that time in the constellation Taurus) could be observed. Observations were made simultaneously in the cities of Sobral Ceará, Brazil, and in São Tomé and Príncipe on the west coast of Africa. The result was considered spectacular news and made the front page of most major newspapers. It made Einstein and his theory of general relativity world-famous. When asked by his assistant what his reaction would have been if general relativity had not been confirmed by Eddington and Dyson in 1919, Einstein famously made the quip: "Then I would feel sorry for the dear Lord. The theory is correct anyway."

The early accuracy, however, was poor. The results were argued by some to have been plagued by systematic error and possibly confirmation bias, although modern reanalysis of the dataset suggests that Eddington's analysis was accurate. The measurement was repeated by a team from the Lick Observatory in the 1922 eclipse, with results that agreed with the 1919 results and has been repeated several times since, most notably in 1953 by Yerkes Observatory astronomers and in 1973 by a team from the University of Texas. Considerable uncertainty remained in these measurements for almost fifty years, until observations started being made at radio frequencies. It was not until the 1960s that it was definitively accepted that the amount of deflection was the full value predicted by general relativity, and not half that number. The Einstein ring is an example of the deflection of light from distant galaxies by more nearby objects.

Gravitational Redshift of Light

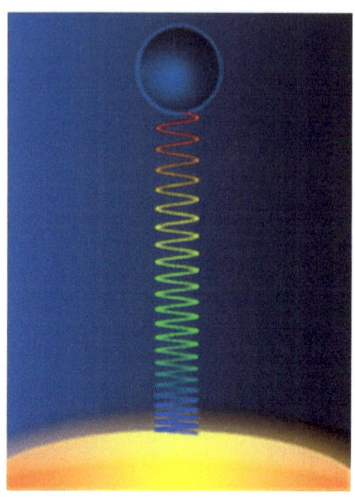

(Above) The gravitational redshift of a light wave as it moves upwards against a gravitational field, caused by the yellow star below.

Einstein predicted the gravitational redshift of light from the equivalence principle in 1907, but it is very difficult to measure by astro-physics rules. Although it was measured by Walter Sydney Adams in 1925, it was only conclusively tested when the Pound-Rebka experiment in 1959 measured the relative redshift of two sources situated at the top and bottom of Harvard University's Jefferson tower using an extremely sensitive phenomenon called the Mössbauer effect. The result was in excellent agreement with general relativity. This was one of the first precision experiments testing general relativity.

Modern Tests

The modern era of testing general relativity was ushered in largely at the impetus of Dicke and Schiff who laid out a framework for testing general relativity. They emphasized the importance not only of the classical tests, but of null experiments, testing for effects which in principle could occur in a theory of gravitation, but do not occur in general relativity. Other important theoretical developments included the inception of alternative theories to general relativity, in particular, scalar-tensor theories such as the Brans-Dicke theory, the parameterized post-Newton formalism in which deviations from general relativity can be quantified; and the framework of the equivalence principle.

Experimentally, new developments in space exploration, electronics and condensed matter physics have made additional precise experiments possible, such as the Pound Rebka experiment, laser interferometry and lunar range finding.

Newtonian Tests of Gravity

Early tests of general relativity were hampered by the lack of viable competitors to the theory: it was not clear what sorts of tests would distinguish it from its competitors. General relativity was the only known relativistic theory of gravity compatible with special relativity and observations. Moreover, it is an extremely simple and elegant theory. This changed with the introduction of the Brans-Dicke theory in 1960. This theory is arguably simpler, as it contains no dimension constants, and is compatible with a version of Mach's principle and Dirac's large numbers hypothesis, two philosophical ideas which have been influential in the history of relativity. Ultimately, this led to the development of the parametrized post-Newtonian formalism by Nordtvedt and Will, which parametrizes, in terms of ten adjustable parameters, all the possible departures from Newton's law of universal gravitation to first order in the velocity of moving objects (*i.e.* to first order in v/c, where v is the velocity of an object and c is the speed of light). This approximation allows the possible deviations from general relativity, for slowly moving objects in weak gravitational fields, to be systematically analyzed. Much effort has been put into constraining the post-Newtonian parameters, and deviations from general relativity are at present severely limited.

The experiments testing gravitational lensing and light time delay limits the same post-Newtonian parameter, the so-called Eddington parameter γ, which is a straightforward parametrization of the amount of deflection of light by a gravitational source. It is equal to one for general relativity, and takes different values in other theories (such as the Brans–Dicke theory). It is the best constrained of the ten post-Newtonian parameters, but there are other experiments designed to constrain the others. Precise observations of the perihelion shift of Mercury constrain other parameters, as do tests of the strong equivalence principle.

One of the goals of the Bepi Colombo Mission is testing the general relativity theory by measuring the parameters gamma and beta of the parametrized post-Newtonian formalism with high accuracy.

Sound of 2 Black Holes Colliding

Sound of 2 Black Holes Colliding

Chapter 2. Artists' Conceptions of Black Holes Colliding

(Above) An artist's conception of two colliding black holes which sent ripples (called Gravitational Waves) the space time fabric of the Universe.

This chapter shows a collection of artists' conceptions of the Gravitational Waves which were found by the Laser Interferometer Gravitational Wave Observatory in Louisiana and in Washington State. For the first time ever, scientists had recorded a gravitational wave signal.

Sound of 2 Black Holes Colliding

Binary black holes radiate a huge amount of orbital energy as gravitational waves.

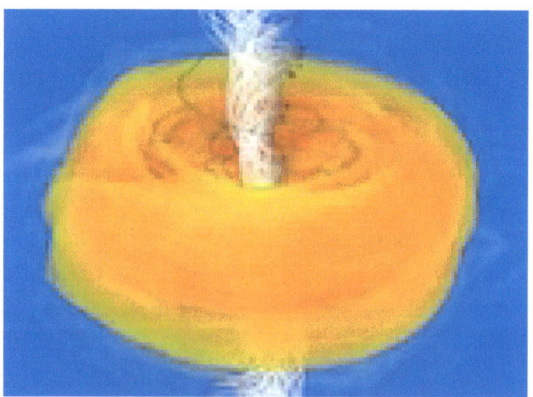

Above, 3D Simulation of spinning black holes.

Sound of 2 Black Holes Colliding

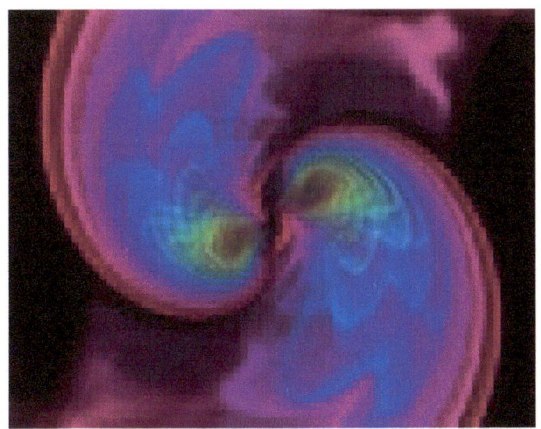

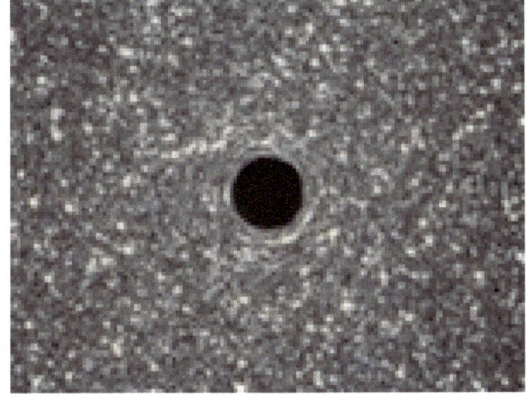

Sound of 2 Black Holes Colliding

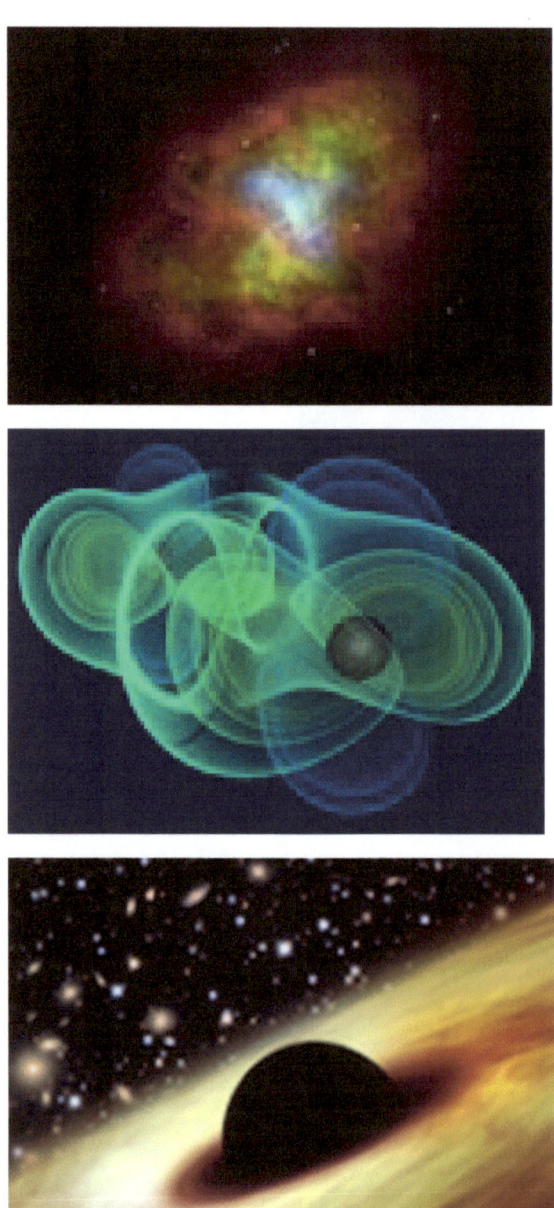

Sound of 2 Black Holes Colliding

Chapter 3. Black Hole Description

Image of a black-hole (above)

A black hole is a place in space where gravity pulls so much that even light can not get out. The gravity is so strong because matter has been squeezed into a tiny space. This can happen when a star is dying.

Because no light can get out, people can't see black holes. They are invisible. Space telescopes with special tools can help find black holes. The special tools can see how stars that are very close to black holes act differently than other stars.

Sound of 2 Black Holes Colliding

How Big Are Black Holes?

Black holes can be big or small. Scientists think the smallest black holes are as small as just one atom. These black holes are very tiny but have the mass of a large mountain. Mass is the amount of matter, or "stuff," in an object.

Another kind of black hole is called "stellar." Its mass can be up to 20 times more than the mass of the sun. There may be many, many stellar mass black holes in Earth's galaxy. Earth's galaxy is called the Milky Way.

The largest black holes are called "supermassive." These black holes have masses that are more than 1 million suns together. Scientists have found proof that every large galaxy contains a supermassive black hole at its center. The supermassive black hole at the center of the Milky Way galaxy is called Sagittarius A. It has a mass equal to about 4 million suns and would fit inside a very large ball that could hold a few million Earths.

Sound of 2 Black Holes Colliding

Image above: depicting a massive black-hole in the middle of the Milky Way: credit: Nasa/JPL-Caltech

How Big Are Black Holes?

Black holes can be big or small. Scientists think the smallest black holes are as small as just one atom. These black holes are very tiny but have the mass of a large mountain. Mass is the amount of matter, or "stuff," in an object.

Another kind of black hole is called "stellar." Its mass can be up to 20 times more than the mass of the sun. There may be many, many stellar mass black holes in Earth's galaxy. Earth's galaxy is called the Milky Way.

The largest black holes are called "supermassive." These black holes have masses that are more than 1 million suns together. Scientists have found proof that every large galaxy contains a supermassive black hole at its center. The supermassive black hole at the center of the Milky Way galaxy is called Sagittarius A. It has a mass equal to about 4 million suns and would fit inside a very large ball that could hold a few million Earths.

How Do Black Holes Form?

Scientists think the smallest black holes formed when the universe began.

Stellar black holes are made when the center of a very big star falls in upon itself, or collapses. When this happens, it causes a supernova. A supernova is an exploding star that blasts part of the star into space.

Scientists think supermassive black holes were made at the same time as the galaxy they are in.

If Black Holes Are "Black," How Do Scientists Know They Are There?

A black hole cannot be seen because strong gravity pulls all of the light into the middle of the black hole. But scientists can see how the strong gravity affects the stars and gas around the black hole. Scientists can study stars to find out if they are flying around, or orbiting, a black hole.

When a black hole and a star are close together, high-energy light is made. This kind of light can not be seen with human eyes. Scientists use satellites and telescopes in space to see the high-energy light.

The theory of general relativity continues to be tested with ever more accurate measurements, for example buy NASA's Gravity Probe Satellite.

Could a Black Hole Destroy Earth?

Black holes do not go around in space eating stars, moons and planets. Earth will not fall into a black hole because no black hole is close enough to the solar system for Earth to do that.

Sound of 2 Black Holes Colliding

Even if a black hole the same mass as the sun were to take the place of the sun, Earth still would not fall in. The black hole would have the same gravity as the sun. Earth and the other planets would orbit the black hole as they orbit the sun now.

The sun will never turn into a black hole. The sun is not a big enough star to make a black hole.

How Is NASA Studying Black Holes?

NASA is using satellites and telescopes that are traveling in space to learn more about black holes. These spacecraft help scientists answer questions about the universe.

Sound of 2 Black Holes Colliding

German-Swiss-American physicist Albert Einstein depicted above expanded on Newton's theories by formulating the theory of General Relativity. Einstein died in 1955 while living at the Caltech Athenaeum. While your author was waiting tables at the Athenaeum to pay for his food while attending Caltech in 1951, he was greatly privileged to have Einstein as a guest.

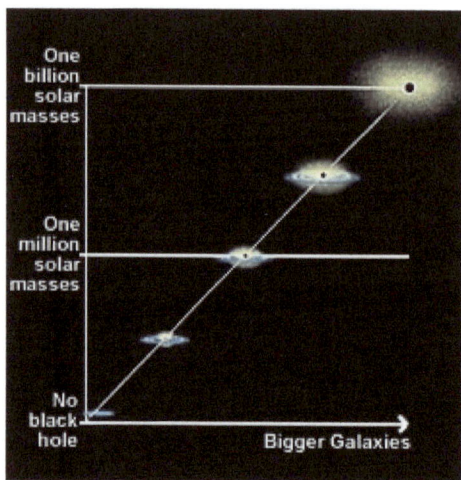

(Above) The chart shows the growth of galaxies from no black hole through a black hole the size of one billion solar masses.

So you may see there is a vast range of black hole sizes throughout the universe.

Note: the book I wrote (*String Theory*) indicates there may by two Universes.

Sound of 2 Black Holes Colliding

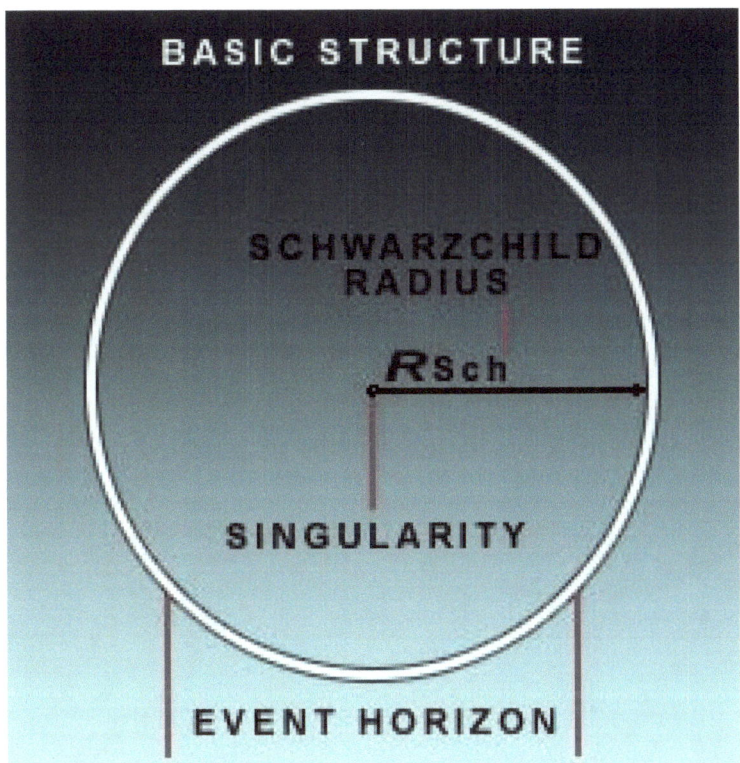

A non-rotating black hole is depicted above. The central point is surrounded by an imaginary sphere called the event horizon. Its size is called the Schwarzschild radius. Karl Schwarzschild (1873 – 1916) first discovered the solutions to the equations that describe non-rotating back holes.

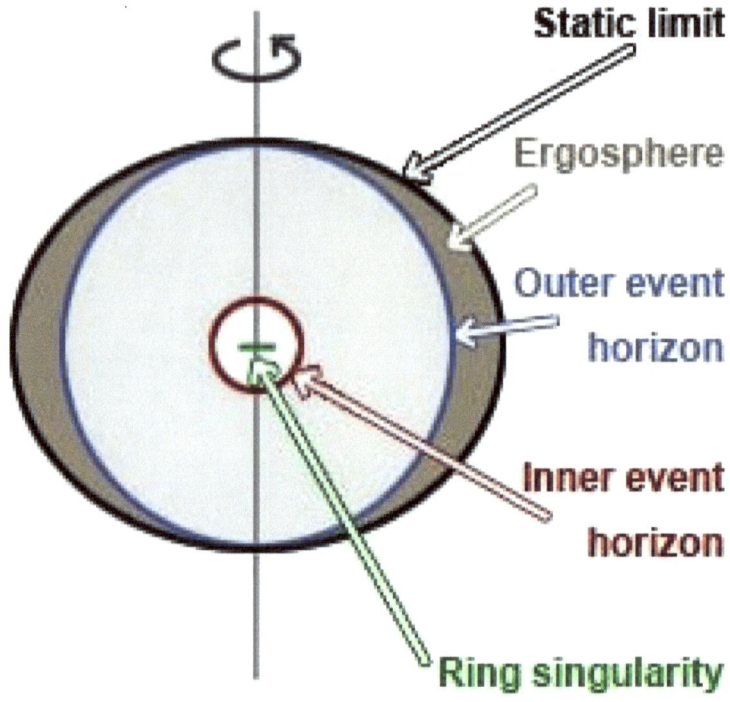

In a spinning black hole, a central ring singularity is surrounded by two event horizons: the ergoshere and the static limit.

Sound of 2 Black Holes Colliding

What Is Inside a Black Hole?

A Wormhole is depicted above which shows a short cut way to connect two points widely separated by the curved space time of the Universe.

Kip Thorne of Caltech studied the possibilities of time travel by utilizing wormholes.

Sound of 2 Black Holes Colliding

Chapter 4. Sound of Two Black Holes Colliding

Below is an artist's concept of two colliding black holes sending ripples through the space-time fabric of the Universe that are called gravitational waves. It is possible for two black holes to collide. Once they come so close that they cannot escape each other's gravity, they will merge to become one bigger black hole.

Sound of 2 Black Holes Colliding

Can We Hear Back Holes Collide?

Many of the stars in the nearby universe are not too different from our Sun: relatively small and long-lived, shining for billions of years. A small fraction, however, are much more massive, and burn their hydrogen "fuel" more rapidly. Over time these stars accumulate a dense core of heavier elements such as carbon and oxygen due to nuclear fusion - See more at: http://www.ligo.org/science/Publication-NINJA2/#sthash.wLtCoTR6.dpuf

The apocalypse is still on, apparently — at least in a galaxy about 3.5 billion light-years from here.

In winter 2015, a team of Caltech astronomers reported that two supermassive black holes appeared to be spiraling together toward a cataclysmic collision that could bring down the curtains in that galaxy.

The evidence was a rhythmic flickering from the galaxy's nucleus, a quasar known as PG 1302-102, which Matthew Graham and his colleagues interpreted as the fatal mating dance of a pair of black holes with a total mass of more than a billion suns. Their merger, the astronomers calculated, could release as much energy as 100 million supernova explosions, mostly in the form of violent ripples in space-time known a gravitational wave that would blow the stars out of that hapless galaxy like leaves off a roof.

In spring 2016, a new analysis of the system by Daniel D'Orazio of Columbia University and his colleagues has added weight to

Sound of 2 Black Holes Colliding

that conclusion. Mr. D'Orazio, a graduate student, and his colleagues, Zoltan Haiman and David Schiminovich, propose that most of the light from the quasar is coming from a vast disk of gas surrounding the smaller of the two black holes.

As the black holes and their attendant disks swing around each other at high speeds, the light from the disk that is coming toward us receives a boost from relativistic effects -- a so-called Doppler boost -- the way a siren grows louder and more high-pitched as it approaches, giving rise to a periodic increase in brightness every five years.

The Columbia astronomers' model predicted that the variation would be two or three times greater in ultraviolet light than in visible light. And that is exactly what they found when they compared archival data from the Hubble Space Telescope and NASA's Galex space telescope to the visible-light data previously analyzed by Dr. Graham's group.

"What's big is that the Doppler boost is inevitable," Dr. Haiman said in an email. Given reasonable assumptions about the masses of the two black holes, their model predicts the right ultraviolet data. "This is rare in 'messy' astronomy," he said, "to have an indisputable clean effect, which explains the data." Follow-up observations of ultraviolet and visible light emissions in the coming years could help clinch the case, the authors said. Their paper was published recently in the journal, *Nature*.

Sound of 2 Black Holes Colliding

A Guide to Black Holes

A black hole is a place of no return — a region in space where the gravitational pull is so strong not even light can escape it.

Even Einstein — whose Theory of General Relativity made it possible to conceive of such a place — thought the concept was too bizarre to exist. But Einstein was wrong. A black hole swallows up everything too close, too slow or too small to fight its gravitational force — even light. With every planet, gas, star or bit of mass consumed, the black hole grows.

(Above) Bright flares are visible near the event horizon of a supermassive black hole at the center of the Milky Way. At the edge of a black hole, its event horizon, is the point of no return. Avoid the event horizon, because that's where the hole pulls in light. And nothing is faster than light. At the event horizon, everything enters the black hole.

If You Fell into a Black Hole, It's Not Clear How You Would Die

Would gravity rip you apart and crush you into the black hole's core? Or would a firewall of energy sizzle you into oblivion? Could some essence of you ever emerge from a black hole? First posited by a group of theorists including Donald Marolf, Ahmed Almheiri, James Sully and Joseph Polchinski in March 2012, the question of how you would die inside a black hole is probably the biggest debate in physics right now. It's called the firewall paradox.

Based on the mathematics in Einstein's 1915 General Theory of Relativity, you would fall through the event horizon unscathed before gravity's force pulled you into a noodle and ultimately crammed you into singularity, the black hole's infinitely dense core.

But Dr. Polchinski and his team pitted Einstein against quantum theory, which posited that the event horizon would become a blazing firewall of energy that would torch your body to smithereens.

Keep both theories, the physicist Stephen Hawking said in January 2014. Black holes aren't what we thought they were. There is no event horizon, and there is no singularity. They're just different.

Sound of 2 Black Holes Colliding

According to Dr. Hawking, at the edge of a black hole, the fourth dimension known as space-time fluctuates like weather, making the crisp edge we assume impossible. Instead, Dr. Hawking's "apparent horizon" would be like a purgatory for light rays attempting to escape a black hole, slowly dissolving and moving inward, but never being pulled into singularity. The event horizon, he says, remains the same, or even shrinks as a black hole slowly leaks energy. Suspended in the apparent zone, you would scramble and leak out into the cosmos as "Hawking radiation."

(Above) Galaxy NGC 1275. Credit NASA.

Sound of 2 Black Holes Colliding

Black Holes Can Sing

In 2003, an international team led by the X-ray astronomer Andrew Fabian discovered the longest, oldest, lowest note in the universe, a black hole's song, using NASA's Chandra X-ray Observatory. Although it is too low and deep for humans to hear, the B flat note, 57 octaves below middle C, appeared as sound waves that moved out from explosive events at the edge of a supermassive black hole in the galaxy NGC 1275.

The notes stayed in the galaxy and never reached us, but we couldn't have heard them anyway. The lowest note the human ear can detect has an oscillation period of one-twentieth of a second. This B flat's period was 10 million years.

The "songs" of black holes may be behind a declining birth rate of stars in the universe. In clusters of galaxies such as Perseus, the home of NGC 1275, the energy these notes carry is thought to keep the gases too hot to condense and form stars.

Black Holes May Control the Size of a Galaxy.

Playing music that keeps the intergalactic clusters too hot for stars might not be the only way black holes help maintain galaxies. Astronomers think that the energy that forms when galactic masses swirl and heat up around a black hole shoots out in X-ray beams that fuel quasars, supermassive black holes that are actively chomping down gas at the centers of distant galaxies.

Sound of 2 Black Holes Colliding

(Above) A big galaxy gobbles a tiny one. Credit: Swinburne, University of Technology, Reuters.

(Above) The Milky Way as visible from the desert southwest of Cairo. Credit: Amr Abdallah-Dalsh, Reuters.

Sound of 2 Black Holes Colliding

Astronomers Have Evidence of Black Holes in Nearly Every Galaxy in the Universe.

Although no black hole is close enough to Earth to pull the planet into its depths, there are so many black holes in the universe that counting them is impossible. Nearly every galaxy, our own Milky Way as well as the 100 billion or so other galaxies visible from Earth, shows signs of a black hole.

Of the billions of stars in the Milky Way, about one in every thousand new stars is massive enough to become a black hole. Our sun isn't. But a star 25 times heavier is. Stellar-mass black holes result from the death of these stars, and can exist anywhere in the galaxy.

Supermassive black holes, a million to a billion times more massive than our sun, exist only in the center of a galaxy. At the center of the Milky Way, 26,000 light-years from Earth, scientists are hoping to make an image of Sagittarius A, which is believed to be our own supermassive black hole, with the mass of four million suns.

Black Holes Are Stellar Tombstones

On July 2, 1967, a network of satellites recorded an explosion of gamma rays coming from outer space. In retrospect, this was one of the first indications that black holes are real. Today, scientists believe the gamma ray burst was the final breath of a dying star and the birth of a stellar-mass black hole.

Sound of 2 Black Holes Colliding

The dramatic transformation starts when a massive star runs out of fuel to power itself. As the star begins to collapse, it explodes. The star's outer layers spew out into space, but the inside implodes, becoming denser and denser, until there is too much matter in too little space. The core succumbs to its own gravitational pull and collapses into itself, in extreme cases forming a black hole.

Theoretically, if you shrunk any mass down into a certain amount of space, it could become a black hole. Our earth would be one if you tried to cram the earth into a pea.

(Above) NASA's Hubble Space Telescope captured a high energy blast, likely a black hole eating, at the center of a galaxy. Credit NASA

Sound of 2 Black Holes Colliding

On March 28, 2011, astronomers detected a long gamma ray burst coming from the center of a galaxy four billion light-years away. This was the first time humans observed what might have been a dormant black star eating a star.

The more a black hole eats, the more it grows. In 2011, scientists discovered one of the biggest black holes ever, more than 300 million light-years away. It weighs enough to have gobbled up 21 billion suns. Scientists want to know if the biggest black holes are the result of two holes merging or one hole eating a lot.

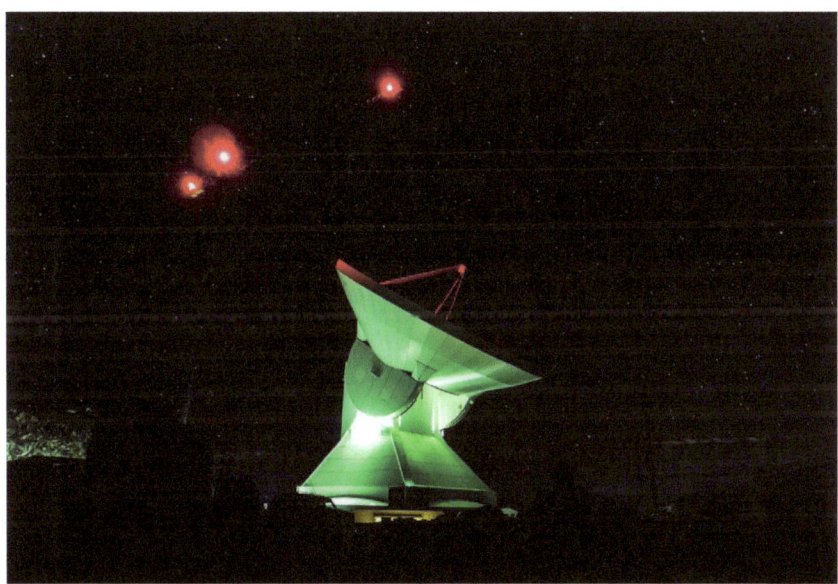

(Above) The Large Millimeter Telescope perched atop the dormant volcano, Sierra Negra, in Mexico is the nerve center for the Event Horizon Telescope, a network of antennas that make up the largest telescope ever. Credit: James D. Lowenthal, Smith College Astronomy Department.

Sound of 2 Black Holes Colliding

To Find the Darkness, Follow The Light

Because light can't escape a black hole, seeing what's inside it is impossible. Getting a picture of a black hole's edge is difficult, and getting a clear picture is an event.

Scientists suspect black holes when their tools detect high-energy radio waves, such as those that may result from a collapsing star, gamma ray burst, supernova or the energy an object might release before reaching the black hole's event horizon. Generally, if there is a lot of energy with a massive core at the center of a galaxy, the core is probably a black hole.

The Event Horizon Telescope, the one Sheperd Doeleman and his colleagues used to try to photograph Sagittarius A* and M87, another black hole, required more than 100 scientists on three continents plus one important crystal used to calibrate atomic clocks. The scientists were at seven telescopes atop six mountains. They synchronized time, pointed their discs at the sky and waited. For the first time ever, scientists may have seen a rough image of a black hole's event horizon.

Sound of 2 Black Holes Colliding

The image on the previous page shows an artist's conception of stars moving in the central regions of a giant elliptical galaxy that harbors a black hole. Credit: Lynette Cook, Gemini Observatory, Nature Magazine, Associated Press.

A Black Hole Is Not Forever

Stephen Hawking suggested that a black hole will evaporate eventually. It would take many times the age of the universe for a black hole to fully evaporate. Like Einstein, at first Dr. Hawking did not believe his own theory. But the numbers were right. Physicists now view his result as the backbone for whatever future theory will bring gravity and quantum theory together.

The Large Hadron Collider in Switzerland

Before the European Organization for Nuclear Research fired up the Large Hadron Collider in 2008, critics worried that smashing together protons in a 17-mile ring underground would create a black hole that would swallow the earth. They forced a safety review.

Some people had worried about Brookhaven National Laboratory's Relativistic Heavy Ion Collider that the center's scientists had squelched nearly 10 years earlier. According to their calculations, ultra-high-energy cosmic rays already penetrated the earth's atmosphere and predicted about 100 tiny black holes on earth every year. If tiny black holes were a problem, Earth would have already collapsed into infinity:

Sound of 2 Black Holes Colliding

In June 2008, a safety review proclaimed the L.H.C. was safe. Experiments commenced, the Higgs Boson was found, and the earth survived after all.

(Above) The muon detector shown above is part of one of the experiments at the Large Hadron Collider, the world's largest and most powerful particle accelerator at the European Organization for Nuclear Research in Geneva. Credit: Fabrice Coffrini, Agence France-Press, Getty Images.

Chapter 5. What Happens When 2 Black Holes Collide

The Black Hole Collision That Reshaped Physics

It was a catastrophic event -- a merger of black holes that violently shook the surrounding space and time, and sent a blast of space time vibrations known as gravitational waves rippling across the Universe at the speed of light on 14 September 2015.

But it was the kind of calamity that physicists on Earth had been waiting for. On September 14, when those ripples swept across

Sound of 2 Black Holes Colliding

the freshly upgraded Laser Interferometer Gravitational Wave (Advanced LIGO), they showed up as spikes in the readings from its two L-shaped detectors in Louisiana and Washington state. For the first time ever, scientists had recorded a gravitational wave signal.

"There it was!" said LIGO team member Daniel Holz, an astrophysicist at the University of Chicago in Illinois. "And it was so strong, and so beautiful, in both detectors." Although the shape of the signal looked familiar from the theory, Holz says, "it's completely different when you see something in the data. It's this transcendental moment".

The signal, formally designated GW150914 after the date of its occurrence and informally known to its discoverers as 'the Event', has justly been hailed as a milestone in physics. It has provided a wealth of evidence for Albert Einstein's 100-year-old General Theory of Relativity, which holds that mass and energy can warp space time, and that gravity is the result of such warping. Stuart Shapiro, a specialist in computer simulations of relativity at the University of Illinois at Urbana–Champaign, calls it "the most significant confirmation of the general theory of relativity since its inception".

But the Event also marked the start of a long-promised era of gravitational-wave astronomy. Detailed analysis of the signal has already yielded insights into the nature of the black holes that merged, and how they formed. With more events such as these -- the LIGO team is analyzing several other candidate events captured during the detectors' four-month run, which ended in

Sound of 2 Black Holes Colliding

January 2016 -- researchers will be able to classify and understand the origins of black holes, just as they are doing with stars.

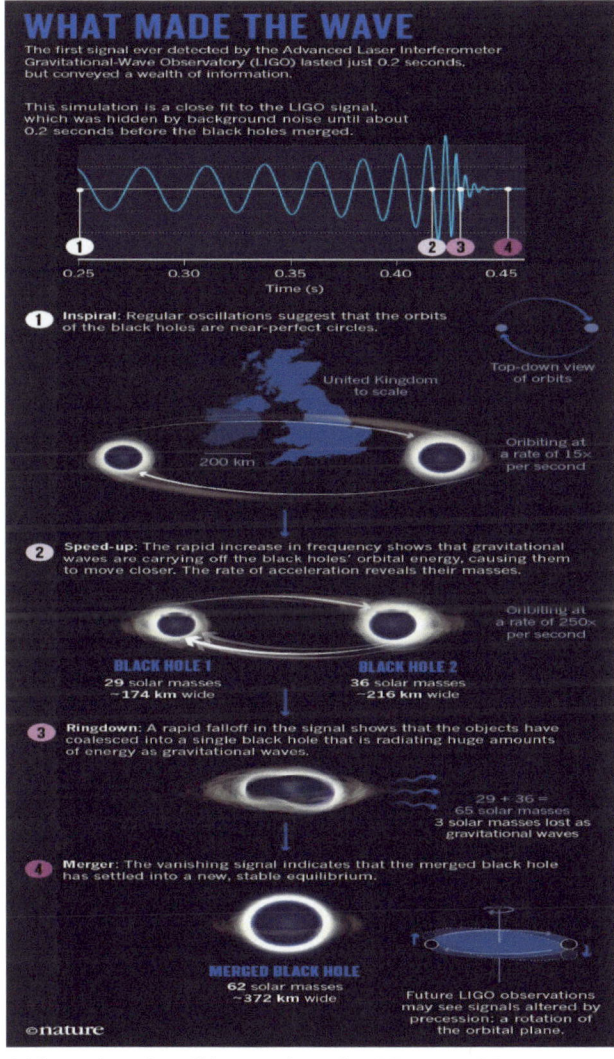

(Above) An illustration in Nature News on March 23, 2016. DOI:10.1038/531428a. Source: Ref. [1]/Nik Spencer/Nature

Sound of 2 Black Holes Colliding

Still more events should appear starting in September 2016, when Advanced LIGO is scheduled to begin joint observations with its European counterpart, the Franco–Italian-led Advanced Virgo facility near Pisa, Italy. (The two collaborations already pool data and publish papers together.) This detector will not only contribute crucial details to events, but could also help astronomers to make cosmological-distance measurements more accurately than before.

"It's going to be a really good ride for the next few years," said Bruce Allen, managing director of the Max Planck Institute for Gravitational Physics in Hanover, Germany.

"The more black holes they see whacking into each other, the more fun it will be," said Roger Penrose, a theoretical physicist and mathematician at the University of Oxford, UK, whose work in the 1960s helped to lay the foundation for the theory of the objects. "Suddenly, we have a new way of looking at the Universe."

A Matter of Energy

Physicists have known for decades that every pair of orbiting bodies is a source of gravitational waves. With each revolution, according to Einstein's equations, the waves will carry away a tiny fraction of their orbital energy. This will cause the objects to move a bit closer together and orbit a little faster. For familiar pairs, such as the Moon and Earth, such energy loss is imperceptible even on timescales of billions of years.

Sound of 2 Black Holes Colliding

But dense objects in very close orbits can lose energy much more quickly. In 1974, radio astronomers Russell Hulse and Joseph Taylor, then of the University of Massachusetts Amherst, found just such a system: a pair of dense neutron stars in orbit around each other. As the years went by, the scientists found that this 'binary pulsar' was losing energy and spiraling inwards exactly as predicted by Einstein's theory.

The two black holes detected by LIGO had probably been losing energy in this way for millions, if not billions, of years before they reached the end. But LIGO did not register the gravitational waves coming from them until 9:50:45 Coordinated Universal Time on September 14, when the wave's frequency rose above some 30 cycles per second (hertz) -- corresponding to 15 full black-hole orbits per second -- and was finally high enough for the detectors to distinguish it from background noise.

But then, in just 0.2 seconds, LIGO watched the signal surge to 250 hertz and suddenly disappear, as the black holes made their final five orbits, reached orbital velocities of half the speed of light and coalesced into a single massive object.

The LIGO and Virgo teams soon went to work extracting as much information as they could from the Event. At the most fundamental level, the signal gave them an existence proof: the fact that the objects came so close to each other before merging meant that they had to be black holes, because ordinary stars would need to be much bigger. "It is, I think, the clearest indication that black holes are really there," said Penrose.

Sound of 2 Black Holes Colliding

The signal also provided researchers with the first empirical test of general relativity beyond regions, including the space around the binary pulsar, where there is comparatively little space time warping. There was no empirical evidence that the theory would keep its validity at the extreme energies of merging black holes, stated Shapiro, but it did.

The signal held a trove of more-detailed information as well. By scrutinizing its shape just before the final cataclysm, the scientists found that it closely approximated a simple sine wave with a steadily increasing frequency and amplitude. According to B. S. Sathyaprakash, a theoretical physicist at Cardiff University, UK, and a senior LIGO researcher, this pattern suggests that the orbits of the black holes were nearly circular, and that LIGO probably had a bird's-eye view of the circles, looking almost straight down on them rather than edge-on.

In addition, the LIGO and Virgo teams were able to use the frequency of the observed wave, along with its rate of acceleration, to estimate the masses of the two black holes: because heavier objects radiate energy in the form of gravitational waves at a faster rate than do lighter objects, their pitch rises more quickly.

By recreating the Event with computer simulations, the scientists calculated that the two black holes weighed about 36 times and 29 times the mass of the Sun, respectively, and that the combined black hole weighed about 62 solar masses. The lost difference, about three Suns' worth, was dispersed as gravitational radiation, much of it during what physicists call the 'ringdown' phase, when

Sound of 2 Black Holes Colliding

the merged black hole was settling into a spherical shape. (For comparison, the most powerful thermonuclear bomb ever detonated converted only about 2 kilograms of matter into energy.---.roughly 10^{30} times less.) The teams also suspect that the final black hole was spinning at perhaps 100 revolutions per second, although the margin of error on that estimate is large.

The inferred masses of the two black holes are also revealing. Each object was presumably the remnant of a very massive star, with the larger star approaching 100 times the mass of the Sun and the smaller one a little less. Thermonuclear reactions are known to convert hydrogen in the cores of such stars into helium much faster than in lighter stars, which leads them to collapse under their own weight only a few million years after they are born. The energy released by this collapse causes an explosion called a type II supernova, which leaves behind a residual core that turns into a neutron star or, if it's massive enough, a black hole.

Scientists say that type II supernovae should not produce black holes much bigger than about 30 solar masses—and both black holes were at the high end of that range. This could mean that the system formed from interstellar gas clouds that were richer in hydrogen and helium than the ones typically found in our Galaxy, and that were poorer in heavy elements -- which astronomers call metals.

Astrophysicists have calculated that stars formed from such low-metallicity clouds should have an easier time forming massive black holes when they explode, explained Gijs Nelemans, an

Sound of 2 Black Holes Colliding

astronomer at Radboud University at Nijmegen in the Netherlands and a member of the Advanced Virgo collaboration. That's because during a supernova explosion, smaller atoms are less likely to be blown away by the blast. Low-metallicity stars thus "lose less mass, so more of it goes into the black hole, for the same initial mass", Nelemans said.

Two by Two

But how did these two black holes end up in a binary system? In a paper published at the same time as the one reporting the discovery, the LIGO and Virgo teams described two commonly accepted scenarios.

The simplest one was that two massive stars were born as a binary-star system, forming from the same interstellar gas cloud like a double-yolked egg, and orbiting each other ever since. (Such binary stars are common in our Galaxy; singletons such as the Sun are the exception, rather than the rule.) After a few million years, one of the stars would have burned out and gone supernova, soon to be followed by the other. The result would be a binary black hole.

The second scenario is that the stars formed independently, but still inside the same dense stellar cluster, perhaps one similar to the globular clusters that orbit the Milky Way. In such a cluster, massive stars would sink towards the center and, through complex interactions with lighter stars, they would form binary systems, possibly long after their transformation into black holes.

Sound of 2 Black Holes Colliding

Simulations made by Simon Portegies Zwart, an astrophysicist at Leiden University in the Netherlands, show that massive stars are more likely to form in dense clusters, where collisions and mergers are more common. He also found that once a binary black-hole system forms, the complex dynamics of the cluster's center would probably kick the pair out at high speed. The binary that Advanced LIGO detected may have wandered away from any galaxy for billions of years before merging, he says.

Although the LIGO and Virgo teams were able to learn a lot from the Event, there is much more that gravitational waves could teach them, even in the case of black-hole mergers. The detectors showed that immediately after the black holes merged, the waves quickly died down as the resulting black hole settled into a symmetrical shape. This is consistent with predictions made by theoretical physicist C. V. Vishveshwara in the early 1970s, a time when "gravitational waves and black holes both belonged to the realm of mythology", he says. "At that time, I had not imagined that it would ever be verified," said Vishveshwara, who is director emeritus of the Jawaharlal Nehru Planetarium in Bangalore, India.

But LIGO saw only just over one cycle of the Event's ringdown waves before the signal became buried once more in the background noise, not yet enough data to provide a rigorous test of Vishveshwara's predictions.

More-stringent tests will be possible if and when LIGO detects black-hole mergers that are larger than this one, or that occur closer to Earth than the Event's estimated distance of 1.3 billion

Sound of 2 Black Holes Colliding

light years, and thus they would give 'louder' waves that stay above the noise for longer.

Alessandra Buonanno, a LIGO theorist and director of the Max Planck Institute for Gravitational Physics in Potsdam-Golm, Germany, said that a more detailed picture of the ringdown stage could reveal how fast the final black hole was rotating, as well as whether its formation gave it a 'natal kick', imparting a high velocity.

In addition, said Sathyaprakash, "we are especially waiting for systems that are much lighter, so they last longer". Such events could include the mergers of lighter binary black holes, of binary neutron stars or of a black hole with a neutron star. Each type would deliver its own signature chirp, and could produce a signal that stays above LIGO's threshold of sensitivity for several minutes or more.

"GW150914 is in some sense a very vanilla system," said Chad Hanna, a LIGO member at Pennsylvania State University in University Park. "It's beautiful, of course, but it doesn't have all the crazy things that one might expect."

Space Artistry

One phenomenon that Sathyaprakash is eager to observe is a 'precession' of the black holes' orbital plane, meaning that their paths trace a kind of 3D rosette. This is a relativistic effect that has no counterpart in Newtonian gravity, and it should produce a characteristic fluctuation in the strength of the gravitational waves. But orbital precession occurs only when two black holes

Sound of 2 Black Holes Colliding

have axes of rotation that point in random directions, and it disappears when the axes are both perpendicular to the orbital plane. The occurrence of a precession could provide clues to how the black holes formed.

It's hard to be sure about that possibility because there are many uncertainties in simulating supernovas. But astrophysicists suspect that parallel spins generally signify that the original two stars were born together out of the same whirling gas cloud. Similarly, they think that random spins result from black holes that formed separately and later fell into orbit around each other. Once the observatories find more mergers, they may be able to determine which type of system occurs more frequently.

Although detecting more events will help LIGO to do lots of science, its interferometers have intrinsic limitations that make it necessary to work together with a worldwide network of similar detectors that are now coming on-line.

First, LIGO's two interferometers are not enough for scientists to determine precisely where the waves came from. The researchers can get some information by comparing the signal's time of arrival at each detector: the difference enables them to calculate the wave's direction relative to an imaginary line drawn between the two. But in the case of the Event, which recorded a difference of 6.9 milliseconds, their calculations limited the field of possibilities merely to a wide strip of southern sky.

Had Virgo been online, the scientists could have narrowed down the direction substantially by comparing the waves' arrival times at three places. With a fourth interferometer (Japan is building an

Sound of 2 Black Holes Colliding

underground one called KAGRA, for Kamioka Gravitational-Wave Detector, and India has its own LIGO in planning), their precision would greatly improve.

Knowing an event's direction will in turn remove one of the biggest uncertainties in determining its distance from Earth. Waves that approach from a direction exactly perpendicular to the detector -- either from above or from below, through Earth -- will be recorded at their actual amplitude, explains Fulvio Ricci, a physicist at the University of Rome La Sapienza and the spokesperson for Virgo. Waves that come from elsewhere in the sky, however, will hit the detector at an angle and produce a somewhat smaller signal, according to a known formula. There are even some blind spots, where a source cannot be seen by a given detector at all.

Determining the direction will therefore reveal the exact amplitude of the waves. By comparing that figure with the waves' amplitude at the source, which the researchers can derive from the shape of the signal, and by knowing how the amplitude decreases with distance, which they get from Einstein's theory, they can then calculate the distance of the source to a much higher precision.

This situation is almost unprecedented: conventionally, astronomical distances need to be estimated by looking at the brightness of known objects in locations that range from the Solar System to distant galaxies. But the measured brightness of those 'standard candles' can be dimmed by stuff in between. Gravitational waves have no such limitation.

Sound of 2 Black Holes Colliding

Raising the Alarm

There is another important reason why scientists are eager to have precise estimates of the waves' provenance. The LIGO and Virgo teams have arranged to give near-real-time alerts of intriguing events to more than 70 teams of conventional astronomers, who will use their optical, radio and space-based telescopes to see whether those events produced any form of electromagnetic radiation. In return, the LIGO and Virgo collaborations will be sifting through data to search for gravitational waves that could have been generated by events, such as supernova explosions, seen by the conventional observatories.

Some 20 teams tried to follow up on the Event, mostly to no avail. NASA's Fermi Gamma-ray Space Telescope did see a possible burst of gamma-rays about 0.4 seconds later, coming from an equally vague but compatible region of the southern sky. But most observers now consider it to be a coincidence. Such gamma rays could, in principle, have been produced when gas orbiting the binary black hole was heated up during the merger, said Vicky Kalogera, a LIGO astrophysicist at Northwestern University in Evanston, Illinois. But "our astrophysical expectation has been that the gas from stars that formed the binary black hole has long dispersed. There shouldn't be any significant gas around", she said.

Going forward, however, matching gravitational waves with electromagnetic ones could usher in a new era of astronomy. In particular, mergers of neutron stars are expected to produce short gamma-ray bursts. Researchers could then measure how far the light from those bursts is shifted towards the red end of the spectrum, which would tell astronomers how fast the stars' host galaxies are receding owing to the expansion of the Universe.

Matching those redshifts to distance measurements calculated from gravitational waves should give estimates of the current rate of cosmic expansion, known as the Hubble constant, that are independent, and potentially more precise, than calculations using current methods. "From the point of view of measuring the Hubble constant, that's our gold-plated source", said Holz.

The LIGO and Virgo teams estimate that they have a 90% chance of finding more events in the data that LIGO has already collected. They are confident that by the time the next run finishes, the event count will be at least 5, growing to perhaps 35 by the end of a run scheduled to start in 2017.

"To be honest," says Holz, "I find it really hard to believe that the Universe is really doing this stuff. But it's not science fiction. It really happened."

After Black Holes Collide, a Puzzling Flash

A surprising burst of light appeared in the sky at the same time as a collision between two black holes. Is the flash just a cosmic coincidence, or will it force astrophysicists to rethink what black holes can do?

Sound of 2 Black Holes Colliding

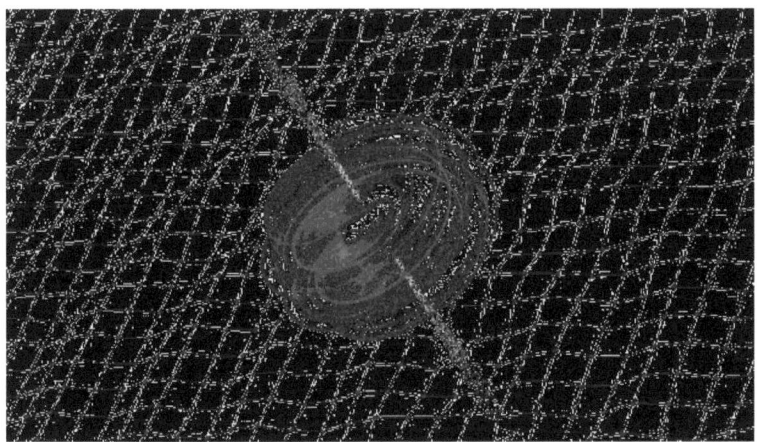

On September 14, 2015, at almost the exact same time that a pair of sprawling gravitational-wave detectors heard the last gasp of a collision between two black holes, another, more perplexing observation took place. Over 500 kilometers above the surface of the Earth, the orbiting Fermi Gamma-Ray Space Telescope logged a passing burst of gamma rays, a high-energy form of light. The signal was so slight that the NASA scientists who run the satellite didn't notice it at first.

LIGO saw a bright event, clear in their data, and we found a little blip in our data that's really only credible because it happened so close in time to the gravitational wave," said Valerie Connaughton, a member of the Fermi team.

On February. 11, 2016, the Fermi researchers posted a paper to the scientific preprint site arxiv.org describing the gamma-ray burst and speculating that it likely originated from the same black-hole merger that produced the gravitational waves observed by LIGO (the Laser Interferometer Gravitational-Wave Observatory). The correlation, which is far from certain, would

Sound of 2 Black Holes Colliding

upend entrenched assumptions in physics. Astrophysicists have long believed that black holes exist in a vacuum, as they tend to swallow up all nearby matter. This absence of matter means it should be impossible for two merging black holes to generate a flash of light.

"If you don't have charged particles, you don't have magnetic fields, and you can't get electromagnetic radiation," said Adam Burrows, an astrophysicist at Princeton University. "It's too clean a system."

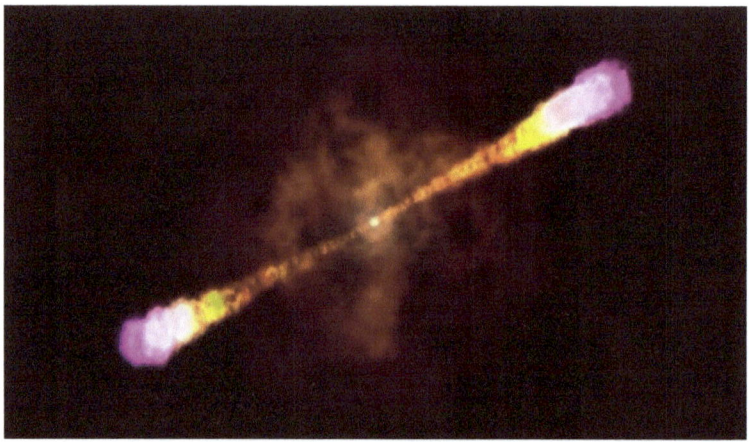

(Above) Gamma-ray bursts most often occur when a massive star collapses to form a black hole. The process blasts two jets of particles outward at nearly the speed of light.

But the gamma-ray burst detected by the Fermi satellite suggested that perhaps the neighborhood around a pair of black holes isn't so empty after all. In the days since the Fermi team released their paper, a number of astrophysicists have hastened to propose theoretical explanations for how matter might persist

Sound of 2 Black Holes Colliding

around black holes in high enough concentrations to generate a gamma-ray burst. These theories involve flights of astrophysical imagination, pulled together in the wake of a historic event, to explain an observation of light that by all accounts should not have been there.

A Cosmic Coincidence?

Gamma rays fall at the very end of the electromagnetic spectrum. Of all the varieties of light, they have the shortest wavelengths, the highest frequency and the most energy — millions of times more energy than ultraviolet light, for example.

It takes extreme conditions to create that much energy, and only two known astrophysical events might do so. One is the collapse of a massive throws off its surrounding envelope of matter and forms violent jets of energy that propel that matter out into space at nearly the speed of light. These are the so-called "long gamma-ray bursts," which account for about 80 percent of all gamma-ray bursts and typically last around 20 seconds.

The second mechanism for producing a gamma-ray burst is the merger of two very compact objects, such as a pair of neutron stars or a neutron star and a black hole. In the case of a star and a black hole, matter from the star forms a ring of material called an accretion disk around the black hole. As the material from the accretion disk falls into the black hole, jets of energy form along the axis of the merger. The result is a "short gamma-ray burst," which typically lasts less than two seconds.

Sound of 2 Black Holes Colliding

Gamma-ray bursts are the great fireworks events of the universe, explosions on a scale that we can hardly imagine. They also provide astrophysicists with a way of seeing hidden cosmic events.

Short gamma-ray bursts allow us to view dark objects. When these objects merge, they produce a violent jet of energetic particles, and the violence appears as a phenomenon that would otherwise look very dark."

On September 14, 2015, Fermi detected a short, transient event that registered as a blip. It was so dim that the team did not even notice it at first. Later, when they learned that LIGO had detected a gravitational wave, they went back through their data to see if Fermi had seen anything interesting at the same time. Using an algorithm developed by Lindy Blackburn, an astronomer at the Harvard-Smithsonian Center for Astrophysics in Cambridge, Massachusetts, and a member of the LIGO team, the Fermi researchers searched for faint blips in their noisy data. That's when they saw it, a burst of gamma rays that arrived 0.4 seconds after the gravitational wave and lasted one second. It had characteristics of a typical short gamma-ray burst that, at its origin, contained 10,000 trillion times the amount of energy the sun produces over that same length of time.

Whether the gamma-ray burst was real, rather than a detection error, and whether, if real, its connection to the LIGO event became a topic of intense debate in the weeks after the Fermi team published their paper.

Sound of 2 Black Holes Colliding

The team has roughly established that the gamma-ray burst came from a 2,000-square-degree area of the sky. Combined with the 600-degree LIGO localization, the arrival direction is reduced to a 200-square-degree patch of sky, supporting the conclusion that the gamma-ray burst and the gravitational waves originated in the same place. The timing of the two events suggests this as well. Fermi has detected blips of this magnitude about once every 10,000 seconds (or about every 2 hours and 47 minutes), making it unlikely, although not impossible, that the near-simultaneous observation of the gamma-ray burst and the gravitational waves was a coincidence.

"It's a low-chance possibility, but it's not impossible that this happened by chance," Connaughton said. "That's why we're circumspect about claiming this is a counterpart to the LIGO event. It's a 'three-sigma' result, not something we'd take to the bank under normal circumstances." In fact, at the same time that Fermi noted the burst, another gamma-ray detector, the European Space Agency's Integral satellite, observed nothing. "From our perspective, it is quite unlikely the event Fermi has detected is related to the gravitational-wave event," said **Carlo Ferrigno**, a member of the Integral team.

More fundamentally, the Fermi team is being cautious about linking the two events because the merger of two black holes is simply not supposed to generate light. "Everything is in its favor, except for physics, which is a problem," Connaughton said.

"To produce a gamma-ray burst you need some conventional matter like an accretion disk around the merging object," said

Sound of 2 Black Holes Colliding

John Ellis, a particle physicist at King's College London. "I think it's pretty clear if you're talking about the merger of neutron stars you'd have that matter. It's not so obvious around black holes."

The accuracy of Fermi's observation will be resolved over time. LIGO will presumably detect more gravitational waves. As it does so, the Fermi team will look for corresponding gamma-ray bursts. If they find them, they'll know they're onto something.

Building Bright Black Holes

In the meantime, astrophysicists have been trying to explain how there could be enough material around a pair of black holes to produce a gamma-ray burst. Bing Zhang, an astrophysicist at the University of Nevada, Las Vegas, has speculated that if one or both of the merging black holes contained a charge, that charge might be sufficient to create a magnetic field that could generate a gamma-ray burst. But according to the general consensus, astrophysical black holes should have no measurable charge.

Another proposal comes from Rosalba Perna, an astrophysicist at Stony Brook University. In a paper posted to arxiv.org on February 16, 2016, she and two colleagues speculated that two massive stars locked together in a binary star system might both die, forming two black holes. As the second massive star in the system dies, debris from its envelope might fall back toward the core and settle into an accretion disk. Then, as a merger begins, the companion black hole would enter the other through this disk, powering a gamma-ray burst.

Sound of 2 Black Holes Colliding

Avi Loeb, the chairman of the astronomy department at Harvard University, has offered a third possibility. In a paper posted to arxiv.org on February 15, 2015, and subsequently accepted for publication in *The Astrophysical Journal Letters*, Loeb described how a pair of black holes might originate simultaneously inside a star 100 times as large as the sun. As he envisioned it, this massive star was originally created when two smaller stars combined. The conditions of that merger could cause the massive star to spin very rapidly. When it eventually began to collapse, the centrifugal force from the spin could cause its core to break into two clumps in a dumbbell configuration, and each clump could form a black hole with the two black holes gravitationally intertwined inside the remnants of the massive star.

The black holes in Loeb's scenario eventually merge, and because the merger takes place inside a massive star, there would be plenty of material around to fuel a gamma-ray burst. In fact, Loeb imagined that as much as a whole solar mass would fall into the newly created black hole per second at the time of the merger.

Loeb's paper is only the beginning of an effort to explain an observation that, if it holds up, would demand a new way of thinking from astrophysicists. A rapidly spinning, supermassive star of the kind at the center of his proposal has never been seen. Additionally, in scenarios where a star has a rapidly rotating inner core, the core doesn't usually split into two dumbbells. Instead, it creates a flattened disc with spiral arms. Over the next year, Loeb and others plan to run computer simulations to determine whether it's possible to generate the conditions described in his paper. Some of Loeb's colleagues are skeptical that his scenario will end up being believable.

"Personally, I think this is a bit of a stretch," said Burrows. "There are a few *tooth fairies* that have been joined here to explain what may be a spurious detection."

Others think that Loeb's paper points the field of astrophysics in the right direction, regardless of whether or not it ends up being correct.

"As always in science when there are important new discoveries, in this case LIGO, there's a time of early speculation where people throw out ideas," said **Volker Bromm**, an astrophysicist at the University of Texas, Austin. "I think Avi's paper is excellent because it focuses people's attention on what needs to be done. It's definitely plausible."

In time, the authenticity of the Fermi detection will become clear. If it does prove accurate, theories will eventually develop that explain how two black holes create a gamma-ray burst. They may resemble the ideas that have been proposed by Zhang, Perna and Loeb, or they may end up looking completely different. What's clear is that post-LIGO, there is a rush to untangle the implications of the post-gravitational-wave world.

Chapter 6. The Particle That Broke a Cosmic Speed Limit

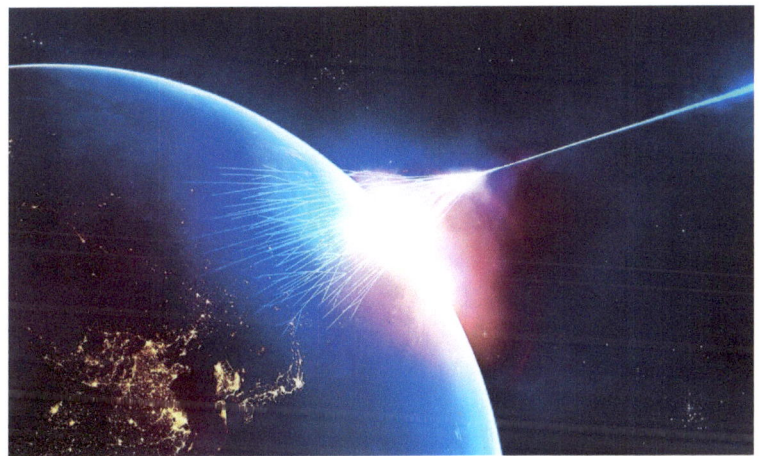

On the night of October 15, 1991, the "Oh-My-God" particle streaked across the Utah sky.

A cosmic ray from space possessed 320 extra-electron volts (EeV) of energy, millions of times more than particles attain at the Large Hadron Collider in Switzerland, the most powerful accelerator ever built by humans. The particle was going so fast that in a year-long race with light, it would have lost by mere thousandths of a hair. Its energy equaled that of a bowling ball dropped on a toe. But bowling balls contain as many atoms as there are stars. "Nobody ever thought you could concentrate so

Sound of 2 Black Holes Colliding

much energy into a single particle before," said David Kieda, an astrophysicist at the University of Utah.

Five or so miles from where it fell, a researcher was working his shift inside an old, rat-infested trailer parked on top of a desert mountain. Earlier, at dusk, Mengzhi "Steven" Luo had switched on the computers for the Fly's Eye detector, an array of dozens of spherical mirrors that dotted the barren ground outside. Each of the mirrors was bolted inside a rotating "can" fashioned from a section of culvert, which faced downward during the day to keep the sun from blowing out its sensors. As darkness fell on a clear and moonless night, Luo rolled the cans up toward the sky. (See picture below.)

"It was a pretty crude experiment," said Kieda, who operated the Fly's Eye with Luo and several others. "But it worked — that was the thing."

Sound of 2 Black Holes Colliding

The faintly glowing contrail of the Oh-My-God particle (as the computer programmer and Autodesk founder John Walker dubbed it in an early Web article) was spotted in the Fly's Eye data the following summer and reported after the group spent an extra year convincing themselves the signal was real. The particle had broken a cosmic speed limit worked out decades earlier by Kenneth Greisen, Georgiy Zatsepin and Vadim Kuzmin, who argued that any particle energized beyond approximately 60 EeV would interact with the background radiation that pervades space, thereby quickly shedding energy and slowing down. This "GZK cutoff" suggested that the Oh-My-God particle must have originated recently and nearby, probably within the local supercluster of galaxies. But an astrophysical accelerator of unimagined size and power would be required to produce such a particle. When scientists looked in the direction from which the particle had come, they could see nothing of the kind.

"It's like you've got a gorilla in your backyard throwing bowling balls at you, but he's invisible," Kieda said.

Where had the Oh-My-God particle come from? How could it possibly exist? Did it really exist? The questions motivated astrophysicists to build bigger, more sophisticated detectors that have since recorded hundreds of thousands more "ultrahigh-energy cosmic rays" with energies above 1 EeV, including a few hundred "trans-GZK" events above the 60 EeV cutoff (though none reaching 320 EeV). In breaking the GZK speed limit, these particles challenged one of the farthest-reaching predictions ever made. It seemed possible that they could offer a window into the laws of physics at otherwise unreachable scales, maybe even connecting particle physics with the evolution of the cosmos as a whole. At the very least, they promised to reveal the workings of extraordinary astrophysical objects that had previously been twinkles in telescope lenses. But over the years, as the particles swept brushstrokes of light across sensors in every direction, instead of painting a telltale pattern that could be matched to, say, the locations of supermassive black holes or colliding galaxies, they created confusion. "It's hard to explain the cosmic-ray data with any particular theory," said **Paul Sommers**, a semiretired astrophysicist at Pennsylvania State University who specializes in ultrahigh-energy cosmic rays. "There are problems with anything you propose."

Only recently, with the discovery of a cosmic ray "hotspot" in the sky, the detection of related high-energy cosmic particles, and a better understanding of physics at more familiar energies, have researchers secured the first footholds in the quest to understand ultrahigh-energy cosmic rays. "We're learning things very rapidly," said **Tim Linden**, a theoretical astrophysicist at the University of Chicago.

Ankle Problems

Thousands of cosmic rays bombard each square foot of Earth's atmosphere every second, and yet they managed to elude discovery until a series of daring hot-air-balloon rides in the early 1910s. As the Austrian physicist, Victor Hess, ascended miles into the atmosphere, he observed that the amount of ionizing radiation increased with altitude. Hess measured this buzz of

Sound of 2 Black Holes Colliding

electrically charged particles even during a solar eclipse, establishing that much of it came from beyond the sun. He received a Nobel Prize in physics for his efforts in 1936.

Cosmic rays, as they became known, arc through Earth's magnetic field from every direction, and with a smooth spread of energies. At sea level, we experienced the low-energy, secondary radiation produced as the cosmic rays crashed through the atmosphere. Most cosmic rays are single protons, the positively charged building blocks of atomic nuclei; most of the rest are heavier nuclei, and a few are electrons. The more energetic a cosmic ray is, the rarer it is. The rarest of all, those that are labeled "ultrahigh-energy" and exceed 1 EeV, strike each square kilometer of the planet only once per century.

Plotting the number of cosmic rays that sprinkle detectors according to their energies produces a downward-sloping line with two bends — the energy spectrum's "knee" and "ankle." These seem to mark transitions to different types of cosmic rays or progressively larger and more powerful sources. The question is, which types, and which sources?

Like many experts, **Karl-Heinz Kampert**, a professor of astrophysics at the University of Wuppertal in Germany and spokesperson for the Pierre Auger Observatory, the world's largest ultrahigh-energy cosmic ray detector, believes cosmic rays are accelerated by something like the sonic booms from supersonic jets, but on grander scales. Shock acceleration, as it's called, "is a fundamental process which you find on any scale in the universe," Kampert said, from solar flares to star explosions

Sound of 2 Black Holes Colliding

(supernovas) to rapidly spinning stars called pulsars to the enormous lobes emanating from mysterious, super-bright galaxies known as active galactic nuclei. All are cases of heated matter (or "plasma") flowing faster than the speed of sound, producing an expanding shock wave that accumulates a crust of protons and other particles. The particles reflect back and forth across the shock waves, trapped between the magnetic field of the plasma and the vacuum of empty space like little balls ping-ponging between table and paddle. A particle gains energy with every bounce. "Then it will escape," Kampert said, "and move through the universe and be detected by an experiment."

Cosmic rays are most likely energized through "shock acceleration," reflecting back and forth across a shock wave that is produced when plasma flows faster than the speed of sound. The stronger and larger the magnetic field of the plasma, the more energy it can impart to a particle. Ultra-high-energy cosmic rays surpass one extra-electron volt (EeV).

Trying to match different shock waves to parts of the cosmic-ray energy spectrum puts astrophysicists on shaky ground, however. They would expect the knee and ankle to mark the highest points to which protons and heavier nuclei (respectively) can be energized in the shock waves of supernovas, the most powerful accelerators in our galaxy. Calculations suggest the protons should max out around 0.001 EeV, and indeed, this aligns with the knee. Heavier nuclei from supernova shock waves are thought to be capable of reaching 0.1 EeV, making this number the expected transition point to more powerful sources of "extragalactic" cosmic rays. These would be shock waves from singular objects that aren't found in the Milky Way or in most other galaxies, and which could well be galaxy-size themselves. However, the measured ankle of the spectrum -- "the only place where it looks like there's a clear transition," Sommers said -- lies around 5 EeV, an order of magnitude past the theoretical maximum for galactic cosmic rays. No one is sure what causes this discrepancy.

Past the ankle, at around 60 EeV, the line dips toward zero, forming a sort of toe. This is probably the GZK cutoff, the point beyond which cosmic rays can only hesitate for so long before losing energy to ambient cosmic microwaves generated by a phase transition in the early universe. The existence of the cutoff, which Kampert calls "the only firm prediction ever made" about cosmic rays, was established in 2007 by the Fly's Eye's successor — the High Resolution Fly's Eye experiment, or HiRes. From there, the energy spectrum reduces to a trickle of trans-GZK cosmic rays, finally ending, at 320 EeV, with a single data point: the Oh-My-God particle.

The presence of the GZK cutoff means that the laws of physics are operating as expected. Rather than disproving those laws, trans-GZK cosmic rays probably do originate nearby (reaching Earth before ambient microwaves sap their energy). But where, and how? For a maddening 20 years, the particles appeared to come from everywhere and nowhere in particular. But finally a hotspot has developed in the Northern Hemisphere. Could this be *the invisible gorilla* hurling bowling balls toward Earth?

Sound of 2 Black Holes Colliding

Getting Hotter

In Utah, a three-hour drive from the site of the original Fly's Eye, its latest descendant sprawls across the desert: a 762–square-kilometer grid of detectors called the Telescope Array. The experiment has been tracking the multi-billion-particle "air showers" produced by ultra-high-energy cosmic rays since 2008. "We've been watching the hotspot increase in statistical significance for several years," said Gordon Thomson, a professor of physics and astronomy at the University of Utah and spokesperson for the Telescope Array.

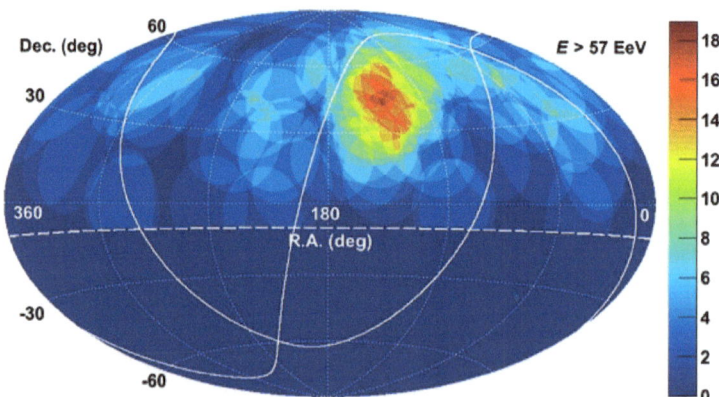

Of the 87 cosmic rays surpassing 57 EeV detected thus far by the Telescope Array, 27 percent come from 6 percent of the sky. The hotspot centers on the constellation Ursa Major.

The hotspot of trans-GZK cosmic rays, which centers on the constellation Ursa Major, was initially too weak to be taken seriously. But recently, it has reached an estimated statistical significance of "four sigma," giving it a 99.994 percent chance of being real. Thomson and his team must reach five-sigma

Sound of 2 Black Holes Colliding

certainty to definitively claim a discovery. (Thomson hopes this will happen in the group's next data analysis, due out in June 2016.) Already, theorists are treating the hotspot as an anchor for their ideas.

"It's really exciting," said Linden. With more data, he explained, the location of the source can be pinpointed within the hotspot (which gets smeared out by the deflection of cosmic rays as they pass through the galaxy's and Earth's magnetic fields). By tracking other types of particles coming from the same spot in the sky, "you have a model of how the source works over many orders of magnitude in energy," he said.

Meanwhile, some of those other particles are slowly piling up in the sensors of the IceCube detector, a cable-infused, cubic-kilometer block of ice buried beneath the South Pole. Since 2014, IceCube has monitored the rare ice tracks of neutrinos, lightweight elementary particles that usually flit right through matter and thus require immense efforts to detect, but which are produced in abundance from physical processes throughout the universe.

Every so often, cosmic neutrinos interact with atoms and produce radiation as they pass through IceCube; their directions of travel trace a new map of the cosmos that can be compared to the maps of ultrahigh-energy cosmic rays and those of light. In 2013, IceCube scientists reported the observation of the first-ever very-high-energy neutrinos — a pair of 0.001-EeV particles nicknamed "Bert" and "Ernie" that might have come from the same sources that yield ultrahigh-energy cosmic rays. Neutrinos have a big advantage over cosmic rays as messengers from the most powerful objects in the universe: Because they are electrically neutral, they move in straight lines. "Since neutrinos travel to us uninhibited from the source, they might be able to open up a new window on the universe," said **Olga Botner** of Uppsala University in Sweden, IceCube's spokesperson.

(Above) At the South Pole, the IceCube Neutrino Observatory is approaching the mystery of ultra-high-energy cosmic rays by hunting related cosmic neutrinos, which interact with atoms every so often while passing through the sensor-infused, cubic-kilometer block of ice

Of the 54 high-energy neutrinos that IceCube has detected as of its latest analysis**, reported in early May 2016,** four originate from the vicinity of the cosmic-ray hotspot. (Neutrinos can enter the detector after traveling through Earth from the northern sky.) This "hint of a correlation," as Linden described it, could be a clue: Cosmic rays take longer to get to Earth than neutrinos, so a common source would have to have been pumping out energetic

Sound of 2 Black Holes Colliding

particles for many years. Short-lived source candidates such as gamma-ray bursts would be ruled out in favor of stable objects, perhaps a star-forming galaxy with a supermassive black hole at its center. "In the next few years we're going to get that many more neutrinos, and we'll see how this correlation plays out," Linden said. For now, though, the correlation is very weak. "I'm not staking my foot in the ground," he said.

Alongside cosmic rays and neutrinos, cosmic "gamma rays" (high-energy photons) will serve as a third messenger in the coming years. They're the subject of several major searches including the HESS (High Energy Stereoscopic System) experiment in Namibia (named in honor of the father of cosmic rays) and VERITAS (Very Energetic Radiation Imaging Telescope Array System) in Arizona, for which Kieda, the former Fly's Eye scientist, now works. The combination of cosmic-ray, neutrino and gamma-ray data should help locate and sharpen astrophysicists' picture of the most powerful accelerators in the universe. The search will organize around the hotspot.

Thomson has his money on threads of galaxies and dark matter called "filaments" that are draped throughout the cosmos and which, at hundreds of millions of light-years long, are among the largest structures in existence. There's a filament in the direction of the hotspot. "It's probably something in the filament," Thomson said. In any case, he added, "we have an idea now of interesting places to look. And all we need to do is collect more data."

Sound of 2 Black Holes Colliding

Draining the Pool

Kampert, of the Pierre Auger Observatory, is approaching the mystery of ultrahigh-energy cosmic rays from a different direction, by asking: What are they?

Some astrophysicists say the Auger Observatory has been "unlucky." Covering 3,000 square kilometers of Argentina grasslands, it collects far more data than the Telescope Array, but it does not see a hotspot in the Southern Hemisphere with anywhere near the prominence as the one in the north. It has detected evidence of a slight concentration of trans-GZK cosmic rays in the sky that overlays an active galactic nucleus called Centaurus-A as well as another filament. But Kampert says Auger might never collect enough data to prove this so-called "warmspot" is real. Still, the dearth of clues is a mystery in itself.

"It's a very rich data set and we don't see anything," said Sommers, who helped design and organize the Auger Observatory. "That's absolutely amazing to me. Back in the 1980s I would have bet good money that if we had the statistics we have now, there would be obvious hotspots and patterns. It makes me really wonder."

Kampert thinks he and his colleagues must simply get smarter about how they look for hotspots, which are surely there; the local region of the universe is not uniformly blanketed by objects capable of accelerating particles to trans-GZK energies. The problem is magnetic deflection, he said. Galactic and extragalactic magnetic fields bend protons five to 10 degrees off-course, and they bend heavier nuclei many times that, depending on the number of protons they contain. Auger's analysis of its air-shower events (which integrates cutting-edge results from particle collisions at the Large Hadron Collider) suggests that the highest-energy cosmic rays tend to be on the heavy side, consisting of carbon or even iron nuclei.

"If at the highest energies we have [heavier nuclei], then your sky is always fuzzy or smeared out," Kampert said. "It would be like doing astronomy from the bottom of a swimming pool."

He and his team hope to update their experiment with the ability to identify the composition of cosmic rays on an event-by-event basis. This will allow them to look for correlations between only the lightest, least deflected particles. "Composition is really the key to understanding the origin of the highest-energy particles," he said.

And the shift toward heavier nuclei at the far end of the cosmic-ray energy spectrum could be a major clue itself. Just as supernovas accelerate protons no further than the "knee" of the spectrum and can propel only heavier nuclei beyond that point, so too might the most powerful astrophysical accelerators in the universe peter out. Scientists could be glimpsing the true edge of the cosmic-ray spectrum: the points where protons, and then helium, carbon and iron, max out. Measuring this falloff will help expose how the giant accelerators work, and favor certain candidates over others.

Theorists still struggle to imagine any of those candidates producing the sprinkle of particles in the 200-EeV range or the

Sound of 2 Black Holes Colliding

Oh-My-God particle at 320, even if they are made of iron. "How you get a [320 EeV] particle is not easy from any theory," Thomson said. "But it was there. It happened."

Even that fact is called into question. Back in the early 1990s, Sommers, who was temporarily working at the University of Utah, helped the Fly's Eye scientists analyze their 320-EeV signal. But although the "Big Event" (as he calls it) was "pretty well measured by the standards of the time," the Fly's Eye hadn't fully transitioned away from being a "monocular" experiment, analogous to one fly's eye rather than two (a second eye was under construction); it lacked the precision and redundancy of later stereoscopic arrays. Sommers said that although no serious reasons for doubting the energy estimate are known, "one must be suspicious of it now. With vastly greater exposure, the more precise, new observatories have failed to detect any particle of such high energy. The flux of particles at energies that high must be so low that it would have been an incredible fluke that the Fly's Eye detected one."

The error bars that went into calculating the Oh-My-God particle's energy might all have been off in the wrong direction at the same time. If so, it was a lucky mistake for the field, motivating new experiments without greatly misleading researchers, since many other trans-GZK particles have followed. And if the Oh-My-God particle was a mistake, well, probably no one will ever know.

Chapter 7. From Einstein's Theory to Gravity's Chirp

Gravitational wave theorists (left to right) Robert Oppenheimer, Roger Penrose, Albert Einstein, Karl Schwarzschild, Arthur Eddington, Kip Thorne and Richard Feynman, whose work helped pave the way for LIGO's big announcement in 14 September 2015. Photo illustration by Olena Shmahalo, Quanta Magazine, Kip Thorne via A.T. Service, Roger Penrose via Festival della Scienza.

"There are no gravitational waves ... " ... "Plane gravitational waves, traveling along the positive X-axis, can therefore be found ... " ... " ... gravitational waves do not exist ... " ... "Do gravitational waves exist?" ... "It turns out that rigorous solutions exist ... "

Sound of 2 Black Holes Colliding

These are the words of Albert Einstein over a period of time. For 20 years he equivocated about gravitational waves, unsure whether these undulations in the fabric of space and time were predicted or ruled out by his revolutionary 1915 theory of general relativity. For all the theory's conceptual elegance -- it revealed gravity to be the effect of curves in "space time" -- its mathematics was enormously complex.

The question was settled once and for all on 14 September 2015, when scientists at the Advanced Laser Interferometer Gravitational Wave Observatory (Advanced LIGO) reported that they had detected gravitational waves emanating from the violent merger of two black holes more than one billion light-years away. Picking up the signal -- a tiny flurry of contractions and expansions in space-time called a "chirp" – required extraordinary technical finesse. But it also took a Century for scientists to determine what, exactly, Einstein's theory predicted: not only that gravitational waves exist, but how they look after crossing the cosmos from a coalescing pair of black holes, inescapably steep sinkholes in space-time whose existence Einstein found even harder to swallow.

Daniel Kennefick, a theoretical physicist at the University of Arkansas, began his career as a graduate student working with LIGO co-founder Kip Thorne to unravel the predictions of general relativity. Fascinated by the contentious history of gravitational-wave research, Kennefick began a sideline as a historian; he is the author of the 2007 book *Traveling at the Speed of Thought: Einstein and the Quest for Gravitational Waves*, and last year he co-authored *An Einstein Encyclopedia*.

Sound of 2 Black Holes Colliding

In discussions before and after the big announcement, Kennefick recounted the journey leading up to it and explained where theorists must go from here. An edited and condensed version of the conversation follows:

Gravitational Waves: Did Merging Black Holes form from a Single Star?

Some people started arguing that even if gravitational waves did exist, it wouldn't be possible to feel them.

In 1955, Nathan Rosen tried to argue that gravitational waves don't carry any energy, so they're just a formal mathematical construct with no real physical meaning. A good way to think about that is, if I'm out in the ocean and there's an enormous ocean swell, I might not even be aware that it's there, because I'll rise up with the wave and then sink back down with it, and so will everything around me. If gravitational waves are like that deep ocean swell, do they really interact with us or do we all just move together up and down in the swell? That was a big debate in the '50s.

How Did That Question Get Resolved?

Rosen's argument was brought up at a conference in 1957 in Chapel Hill, N.C., and very fortunately a man named Felix Pirani, who sadly has recently passed away, came to the conference. He had decided to look at how general relativity works, using a very practical approach that got around this whole problem of the coordinate system, and he showed that the waves would move particles back and forth as they pass by.

Richard Feynman heard Pirani's talk and said, in essence, "Well, since we know that the particles move, all we have to do is imagine a stick, and on the stick we can put some beads. As the wave passes by, the beads will move back and forth, but the stick will stay rigid because the electromagnetic forces in the stick will try to keep the atoms and electrons in the same positions as they were previously. So the beads will drag against the stick, and the friction will produce energy. And the energy must have come from the gravitational wave. So I conclude that the wave has energy." So this famous "sticky bead" thought experiment convinced a lot of people that there wasn't any reason for the skepticism that Rosen had advanced. And then people like Joe Weber started trying to detect gravitational waves shortly after.

But People Still Didn't Know Whether There Would Be Any Astrophysical Sources of Gravitational Waves Strong Enough to Detect, Right?

Right. Einstein wrote that it was unlikely that anyone would ever find a system whose behavior would be measurably influenced by gravitational waves. He was pointing out that the waves from a typical binary star system would carry away so little energy, we would never even notice that the system had changed - and that is true. The reason we can see it from the two black holes is that they are closer together than two stars could ever be. The black holes are so tiny and yet so massive that they can be close enough together to move around each other very, very rapidly. Since

Sound of 2 Black Holes Colliding

Einstein didn't believe in the existence of black holes, he just couldn't conceive of a system that could behave in such a way that you would be able to see the gravitational waves.

Karl Schwarzschild found the black-hole solution to Einstein's equations in 1916, the same year Einstein predicted gravitational waves. Why didn't Einstein believe in black holes after that?

Black holes themselves have a very controversial and complex history, and LIGO's detection was the first really complete proof of the existence of black holes. In 1916 Einstein thought Schwarzschild had just discovered a physical simplification: Just as one would treat the Earth as a point mass [with its mass concentrated to a point] for simplicity, they thought the "Schwarzschild solution" — what we now call a black hole — treated the sun as a point mass just for convenience. They didn't think it would ever be a real thing, where you would have the mass concentrated to a point. They thought that was impossible, outrageous. By the 1930s it was beginning to dawn on people, "You know, it's not entirely clear to us that the theory prevents that from happening." Gradually, people like Robert Oppenheimer, the famous director of the Los Alamos Laboratory for the Manhattan Project, began to show that it was possible for a star to collapse into itself until it actually created something that really did look like the Schwarzschild solution. And that work was taken up in the 1960s by John Wheeler's group, of which Kip Thorne was one of the students, and they and others developed the theory of black holes.

Sound of 2 Black Holes Colliding

How did people then figure out what the gravitational waves produced by merging black holes would look like on Earth?

A key problem was imposing the condition that there are no waves coming into the binary black hole system from infinitely far away, only waves going out to infinity. But that's actually very hard to do, because you usually need a completely different mathematical formalism to describe the very distant gravitational field, at "infinity" or out here at Earth, that you need to describe the black holes themselves. People would try to do this calculation in the 1950s and '60s and they would get wrong answers. In some cases, they would get an answer that the black holes were gaining energy rather than losing it, because they made a mistake and had incoming waves bringing energy in from infinitely far away. So what happened in the course of the 1960s was that people like Roger Penrose, the great English relativist, did research on the structure of space-time. And Penrose discovered that there's more than one infinity at the edge of space and time, and you have to pick the right infinity on which to impose your conditions. And then other people introduced techniques from fluid dynamics. These are just examples of many different conceptual and formulaic breakthroughs that had to be made.

And then the next step was predicting the particular signals that LIGO's detectors might pick up.

At one of my very first group meetings in Kip's group as a young student, this was 1991 or so, he came in with a big sheet of paper, and he had typed up everything that needed to be done on the

Sound of 2 Black Holes Colliding

theory side if LIGO was going to work. Because the whole reason you can detect the signal is that it has this characteristic sweep, and you filter the data against it. But you can only filter if you know what the signal looks like, and since you've never seen it before, you can only know what it looks like if the theorists tell you. And so Kip said, 'I want everybody in the group to work on this.' And that's what we did.

You'd like to have a prediction of the waveform from the beginning of where LIGO could conceivably see the signal to the final stage where the black hole had settled back down again and was not emitting any more waves. But there's no single method that can give you the whole thing. For the first stage, you can use approximation methods that were already around at that time, but it was realized that several orders of magnitude more levels of approximation would be needed, and this was very daunting. And then when the black holes are merging, the gravity is insanely strong, and so you need numerical methods, where you do the calculation on a supercomputer. There were a whole bunch of groups who were trying to do that, and they were confronted with serious challenges. They couldn't evolve the two black holes over more than a tiny amount of time, which wouldn't help at all. And so a few years ago, they basically decided, "We just don't have a choice. We'll keep changing our coordinate systems until we find something that works that doesn't crash on us." And a guy called **Frans Pretorius** found a way to do it, and the methods took off from there.

There's this hope that LIGO will "open up a new window on the universe" by detecting gravitational waves from previously

Sound of 2 Black Holes Colliding

unknown astrophysical objects. Considering the effort that went into recognizing the signal from a black-hole merger, how will we be able to see the unexpected?

Yes, the real excitement would be to find something we didn't expect. One possibility is that the unexpected might help us out by being a very large signal. Our hopes for that have been dampened somewhat, because the original LIGO was online for quite a while and if the signal were very large it might have seen it. It does look like the unexpected is not going to be easy, so how do we dig the signal out of the noise?

One answer is that there are certain kinds of techniques that people have been looking at where you don't commit yourself to knowing precisely what the signal looks like, but you just look for certain kinds of regularities; for instance, maybe this unexpected signal is at least a periodic signal. And LIGO is certainly doing that. They even have an "Einstein@Home" project, where they'll send a piece of LIGO data to your home computer if you sign up for this, and your computer will help look for simple things like that. Another approach is to use machine learning to try to teach machines to look for signals. You start with what you know, but there is some hope that over time these techniques might grow and develop to where they become sufficiently flexible to catch things that aren't what you expect.

A Little Philosophizing About the Above Conversation by One of the Participants

I am struck by the collective nature of the endeavor. It had to be a collaborative effort; each step was sufficiently difficult that it had to link to the next step. And collective efforts come with vitriol and disputes. People shouted at each other. But the finer qualities of human nature won out. People got over their anger. Einstein got over his anger. People admitted they were wrong. And eventually, as a community, we got there.

Chapter 8. Listen to the Collision of Two Black Holes. Einstein was right.

(Above) A frame from a simulation of the merger of two black holes and the resulting emission of gravitational radiation (colored fields, which represent a component of the curvature of space-time). The yellow areas near the black holes do not correspond to physical structures but generally indicate where the strong non-linear gravitational-field interactions are in play.

Gravity. You know it as what holds things together; it holds us to the Earth, holds the Earth in orbit around the sun. But what exactly *is* gravity?

Sound of 2 Black Holes Colliding

More than 300 years ago, Isaac Newton said that any two objects that have mas are attracted to each other and held together by a force.

And that force is what Newton called gravity. He could calculate it, but he couldn't explain where gravity came from.

And, says Priya Natarajan, an astrophysicist at Yale, that's how things stayed for more than 200 years. Until Albert Einstein showed up.

"Einstein thought about gravity in a fundamentally different way," Natarajan says.

Here's where things get fun. I want you to close your eyes. Imagine a large rubber sheet, like a trampoline. Now drop a large metal ball onto it. The ball causes the sheet to bend beneath it, forming a dimple. The bigger the ball, the bigger the dimple.

OK, scale it up. Now the ball is the Sun, and it's sitting not on a rubber sheet, but rather in a four-dimensional fabric, what Einstein called space-time.

"Masses like the Earth or the Sun bend the space-time around them, and by bending the space-time around them, they effectively attract nearby objects," says MIT physicist Matthew Evans.

He says that's how Einstein understood gravity — gravity is what happens when objects bend space-time.

Sound of 2 Black Holes Colliding

Einstein saw it as one of the four fundamental forces of our universe — including electromagnetism and the two forces at work inside atoms. But even though gravity is the force that is most obvious in our lives, it turns out that gravity is the weakest. So to see its effect, you need something dramatic, something that creates massive ripples in space-time, like waves moving out from a rock dropped in a pond. In fact, Einstein called these gravitational waves.

In September 2015, physicists started recording data from the new Advanced LIGO detectors, a pair of ultra-precise observatories waiting for the slightest perturbation that could suggest the existence of gravitational waves, the ripples in spacetime predicted by Einstein almost exactly 100 years ago. Then, after just 16 days, they found it.

The team spent the next several months double-checking every angle to make sure that this was, in fact, the result they were looking for. It was.

"Ladies and gentlemen, we have detected gravitational waves," said David Reitze, executive director of LIGO Laboratory, in a press conference. "We did it."

The article on the next page is one of many based on the information given out by David Reitze. It explains why this book is called the **SOUND** of 2 Black Holes Colliding.

Sound of 2 Black Holes Colliding

Image of colliding black holes above

What Are Gravitational Waves, and Why Should You Care?

Gravitational waves are an important prediction of Einstein's general theory of relativity, published a century ago. Because this theory changed the way we understand the nature of space, time, and gravity, it also fundamentally changed our perception of how we fit into the universe, both as a species and as individuals. Last week's announcement of the first direct detection of gravitational waves reaffirms Einstein's incredible achievement, further demonstrating the power that we have as human beings when we put our minds to constructive rather than destructive purposes.

Einstein's theory of relativity tells us that space and time are intertwined as a four-dimensional *spacetime*. Spacetime has a structure that can vary from place to place (and time to time), much as a two-dimensional surface — like a rubber sheet or the surface of a pond — can have varying bumps, dips, and ripples.

Sound of 2 Black Holes Colliding

The structure of spacetime is shaped by the gravity of the objects within it, so if these objects undergo certain types of movement or change, they can cause a change in the structure of spacetime around them. According to General Relativity, this change then propagates outward through the universe like ripples on a pond, and it is these spacetime ripples that we call *gravitational waves*.

What Do Gravitational Waves Do That Allows Them to Be Detected

Gravitational waves travel outward from their source at the speed of light (a prediction confirmed by the new discovery), slightly distorting space as they pass through it, with the key word being "slightly." The changes predicted to occur as gravitational waves pass by are so small that Einstein himself doubted that we'd ever be able to detect them. But advances in science and technology proved Einstein too pessimistic in this case, since they have now been observed with the detectors known as LIGO, short for *Laser Interferometer Gravitational-Wave Observatory*.

Gravitational waves are not a form of sound. However, like sound waves, they cause vibrations in the material they pass through, and it turns out that the frequencies (that is, the number of vibrations each second) of many gravitational waves happen to be the same as the frequencies of sound waves audible to the human hear. For that reason, it is possible to create sound waves with the same frequencies as the gravitational waves, and in this case the sound happens to be somewhat like a bird's chirp. Bottom line: Gravitational waves do *not* produce sound, but we can artificially create sounds with the same wave pattern and that

Sound of 2 Black Holes Colliding

can be useful when trying to interpret the gravitational wave signal.

Sound of 2 Black Holes Colliding

Chapter 9. Why Black Holes Have a Hard Time Getting Together

The universe's greatest sinkholes have no trouble swallowing anything—except themselves.

It begins like a classic romance: Two black holes meet. The attraction is practically instant. They dance around each other, swirling closer and closer.

Image Above of 2 Black Holes rotating next to eachother

"Our whole picture of the formation of the structure of the universe involves this hierarchical process of small galaxies merging together to form large galaxies, and large galaxies

Sound of 2 Black Holes Colliding

merging to become even larger galaxies," said physicist Robert Owen, who studies black-hole collisions at Oberlin College, in Ohio, as part of the "Simulating Extreme Spacetimes" collaboration. Each merger takes hundreds of millions of years or more—too long to see in action—but theorists can use simulations to recreate the entire affair in computer code.

And here's where the theory and reality run into each other. When physicists run their simulations, the two central black holes in a pair of colliding galaxies get stuck. Rarely, if ever, do black holes crash head-on. Instead, because they are typically traveling along separate, unaligned paths when they meet, their conserved angular momentum causes them to spiral toward each other. They corkscrew ever closer, captives of their mutual attraction, until they are orbiting at arm's length, on the order of three light-years, or one parsec, apart. Then, like bashful lovers, they go no farther.

If physicists' story about the formation of the universe is correct, such paired black holes should eventually collide and consume each other, becoming one. But to do this, they must somehow lose enough energy to resume their inward spiral past the final parsec. Once they get very close, just billions of miles apart (about 0.001 parsec) — General Relativity says that they will jettison the last of their orbital momentum in a great crescendo of gravitational waves, ripples in spacetime that ring out from a gravitational disturbance. This final outburst of energy plunges the black holes together, finishing the job in a matter of hours, days, or years, depending on how massive the black holes are.

Sound of 2 Black Holes Colliding

What drives this fatal embrace? The question, known as the "final parsec problem," isn't just a matter of curiosity. The answer could change our understanding of how the universe built up its elaborate structure, and of the nature of gravity itself. Which is why, as physicists tinker with their simulations, astronomers are searching the skies for clues to how black holes solve the final parsec problem in the wild—if they do at all.

The image above shows the "Galactic Lovers." Astronomers have observed galaxy pairs like this one, known as NGC 4676 or "The Mice" in varioius stages of collision.

Over the past 30 years, astronomers have collected snapshots of hundreds of galaxies with dual supermassive black holes in various stages of collision. But even the most intimate portraits don't reveal pairs circling nearer than a few thousand parsecs. "Looking for ones which are much closer to merger, on the parsec scale or smaller, is much harder," says computational scientist Matthew Graham at the California Institute of Technology. Even the biggest telescopes on Earth can't zoom in enough to resolve an image of two black holes in such a tight orbit.

Sound of 2 Black Holes Colliding

So Graham and his colleagues are instead searching by an indirect route: flickering quasar light. Quasars are tremendously bright cores of massive, ancient galaxies. As matter swirls toward the supermassive black holes at their centers, it accumulates into a disc whose angular momentum converts some of this mass into radiation that outshines the galaxy itself. Because gas and dust don't flow into the disc in a smooth stream, quasar light varies, typically in a random pattern.

But in late 2013 came a quasar that "stood out like a sore thumb," Graham said. Using 10 years' worth of data from a collaboration called the Catalina Real-Time Transient Survey, he and his colleagues picked up a strangely predictable signal: Some 3.5 billion light-years from Earth, quasar "PG 1302-102" appeared to be getting steadily brighter and dimmer every five and a half years, as if someone were slowly toggling some cosmic dimmer switch.

These dense, spinning copses of exploded stars dot the cosmos like buoys on an ocean. Sweeping beams of radio waves rush past Earth with atomic clock accuracy. What could be producing this cycle? "We came up with four or five different physical scenarios," Graham said. The revolutions of a second supermassive black hole, for instance, could be routinely redirecting the quasar's radiation jets like searchlight beams. Or perhaps this extra black hole was contorting the disc of whirling matter, thus brightening and dimming the quasar on a regular schedule. All of the researchers' explanations had one thing in common: They made sense only if the black hole at the center of PG 1302- 102 was actually two black holes.

Sound of 2 Black Holes Colliding

(Above) Exploded Star blooms like a cosmic flower.

If there really is a black-hole binary at the center of PG 1302-102, Graham and his team estimate their separation at just 0.01 parsec. Another analysis, by a team at Columbia University, puts the pair even closer, at 0.001 parsec, or roughly the diameter of our solar system—about the point at which the black holes should be shedding gravitational waves like layers of clothing, plunging them into one another's arms. Either way, if researchers are reading the signals from PG 1302-102 correctly, the moral is the same: Nature has solved the final parsec problem.

Sound of 2 Black Holes Colliding

Graham and other researchers have so far identified more than 100 quasars in the Catalina data set that they think could contain black-hole binaries, all easily within the final parsec. If they can confirm their suspicion, these candidates could give them a peek at the grand finale of the collision saga, which nature has kept so well-hidden.

The big break in the final parsec problem, however (the revelation of how black holes unlock themselves from a stable orbit to complete their union) may come from looking at the universe in an entirely new way. "We're really just fumbling around with electromagnetic waves," Owen said, describing efforts to find tight black-hole binaries using traditional telescopes. Theoretically, a black-hole merger should release 100 million times as much energy as a supernova explosion, but all of that energy comes in the form of gravitational waves, not light. "We're trying to hear with our eyes. It's like inferring that a drum is oscillating just by looking at it, without being able to hear it."

Observing black-hole collisions via gravitational waves could give astronomers a much clearer view. "The light coming from the centers of galaxies is often absorbed, re-emitted, or scattered by clouds of gas and dust," producing a dim and distorted picture, explained Chiara Mingarelli, an astrophysicist at Caltech and the Max Planck Institute for Radio Astronomy. "[Gravitational] ripples don't care if there's gas and dust. They move through it, undisturbed. This is the fabric of spacetime itself moving."

Sound of 2 Black Holes Colliding

Spotting these ripples, however, won't be easy: Gravitational-wave astronomy is a fledgling science. What's more, state-of-the-art laser-based observatories such as LIGO aren't sensitive to the slowly oscillating waves that astronomers suspect are pumping out of intimate black-hole binaries like PG 1302-102.

Researchers hope instead to pick up these disturbances using "telescopes" provided by nature: millisecond pulsars. By monitoring the tick-tock of dozens of millisecond pulsars (a "pulsar timing array") in our own galaxy, the Milky Way, astronomers can look for telltale deviations that reveal the surge of gravitational waves from two black holes crossing through the final parsec in a distant galaxy.

The spectral signature of these waves (from rapid flutters to slow swells, and everything in between) would provide data against which physicists can test new or revised models of the unification process. "Pulsar timing arrays are the only instrument we have to tell us what's happening on this last-parsec scale. What's really driving the final stages of binary black-hole mergers," said Joseph Simon, a graduate student studying these collisions at the University of Wisconsin-Milwaukee.

The absence of gravitational waves could also provide an important clue. After almost a decade of timekeeping, Simon said, pulsar timing arrays "are finally sensitive enough that even non-detection tells us about what's happening." The fact that these arrays haven't yet picked up the scent of gravitational waves could mean that theorists' understanding of what happens to colliding black holes once they cross the final parsec isn't

Sound of 2 Black Holes Colliding

quite right. Rather than erupt as gravitational radiation, some of the energy lost in that final plunge may instead bleed away through some yet unknown interaction with nearby stars and gas. Maybe the black holes fling away some stars that veer toward them, for instance. Or maybe their gravitational pull torques the disc of dust and gas around them. If physicists can work out this energy-sapping mechanism, it might explain how merging black holes cross the final parsec in the first place.

Their calculations will take them to the edges of Einstein's predictions. "We talk about general relativity like it is an extremely well-confirmed theory, and by some measures it is the most precisely confirmed theory in physics," Owen said. But scientists have never tested it in extreme gravitational events, such as a black hole merger, where physics dramatically diverges from the laws laid out by Isaac Newton more than three centuries ago; where familiar concepts like energy, momentum, and mass lose their meaning. If it turns out that gravitational outbursts from black-hole unions are indeed weaker than General Relativity says they should be, it may be time for a tweak.

Ultimately, completing the black-hole love story will tell us what kind of ride we're on here on Earth, whether we're rolling along on a deluge of gravitational waves, or just a trickle. "This really is the difference between a very calm extragalactic sea of space time and a very violent sea of space time," Owen said.

Appendix A. Glossary

Accretion Disk

The disk around a black hole into which falling gas settles before being swallowed. Friction and magnetic fields in the disk cause the gas to heat and the resulting X-rays can sometimes be observed.

Active Black Hole

A black hole that is consuming large amounts of material and which produces bright X-rays and radio waves in the process.

Andromeda Galaxy

A nearby spiral galaxy which is similar to our own galaxy in size and shape.

Binary Star System

2 stars that orbit each other bound together by gravity.

Sound of 2 Black Holes Colliding

Black Hole

An object so compact that it's gravitational force prevents anything from escaping to the outer-world.

Chandra X-Ray Observation

Made by a powerful x-ray telescope in orbit around the Earth in May 2016 named after a Nobel Prize winning astronomer.

Constellation

A pattern of stars grouped together seen as a cluster in the night sky.

Dust

Material spread between the stars similar in size to the particles in cigarette smoke.

Electromagnetic Radiation

A technical term for light which propagates through space as an object in which gravity is balanced by the pressure exerted by densely packed electrons. The object is about the size of the Earth and about as massive as the Sun. It is produced when a low-mass star dies after it runs out of nuclear fuel.

Wormhole

A mysterious solution of the equations of general relativity that resembles a black hole but has no event horizon. May or may not exist in our Universe.

X-rays

Radiation with much smaller wavelength and much higher energy than visible light. We encounter X-rays in our daily lives, for example, in medical imaging devices and security screening machines.

Sound of 2 Black Holes Colliding

www.ingramcontent.com/pod-product-compliance
Lightning Source LLC
Chambersburg PA
CBHW040806200526
45159CB00022B/32